ROYAL NAVY SUBMARINE

1945 to 1973 (A-class – HMS *Alliance*)

COVER CUTAWAY:

HMS *Alliance* as built, 1945. *(Ian Moores)*

© Peter Goodwin 2015

All rights reserved. No part of this publication may be reproduced or stored in a retrieval system or transmitted, in any form or by any means, electronic, mechanical, photocopying, recording or otherwise, without prior permission in writing from Haynes Publishing.

First published in April 2015

A catalogue record for this book is available from the British Library.

ISBN 978 0 85733 770 2

Published by Haynes Publishing,
Sparkford, Yeovil,
Somerset BA22 7JJ, UK.
Tel: 01963 442030 Fax: 01963 440001
Int. tel: +44 1963 442030
Int. fax: +44 1963 440001
E-mail: sales@haynes.co.uk
Website: www.haynes.co.uk

Haynes North America Inc.,
861 Lawrence Drive, Newbury Park,
California 91320, USA.

Printed in the USA by Odcombe Press LP,
1299 Bridgestone Parkway, La Vergne,
TN 37086.

Acknowledgements

This book could not have been written without the enthusiastic support of the following individuals from the Royal Navy Submarine Museum: Commodore Chris Munns RN, Bob Mealings, George Malcolmson, Deborah Turner-West, Alexandra Geary, Bill Sainsbury, and submarine guide Bill Handyside. I would also like to thank Jonathan Falconer at Haynes Publishing for his diligent editing and guidance through what has proved to be a somewhat complex subject. (It could be agreed that Jonathan has passed his 'Part III training' in submariner 'speak'.) My thanks also go to Wynn Davis for his insights into *Alliance's* restoration; gunnery historian Lt Cdr Brian Witts RN (Ret'd), Curator, HMS *Excellent*; ex-submariner and friend Kevin 'Pony' Moore, with whom I briefly served in *Acheron*, for his reminiscences of cooking in A-boats. Finally, and most importantly, I thank my wife Katy who, as on all previous occasions, has given her unconditional support as a writer's partner. As for me, an ex-submariner, researching and writing this book has been a wonderfully nostalgic experience.

Peter Goodwin
January 2015.

Author Peter Goodwin in the engine room of HMS Alliance. *(Katy Ball)*

ROYAL NAVY SUBMARINE

1945 to 1973 (A-class – HMS *Alliance*)

Owners' Workshop Manual

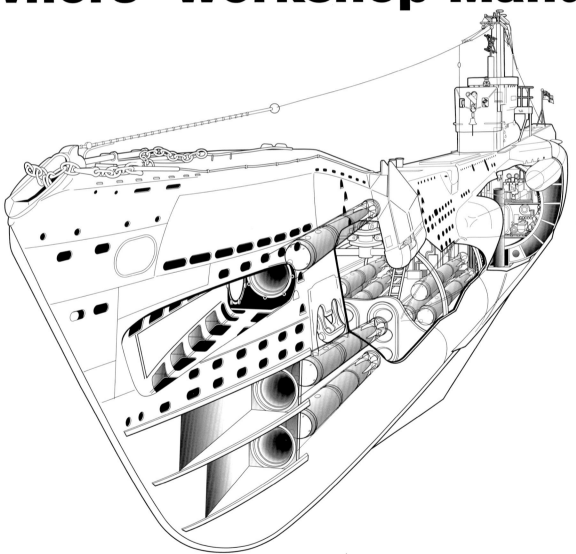

An insight into the design, construction, operation and restoration of a Cold War diesel-electric submarine

Peter Goodwin

Contents

6	Introduction
Preserving *Alliance* for future generations	8
The Royal Navy Submarine Museum	9

10	British submarine development, 1901–45
Birth of the Royal Navy Submarine Service	12
The first ten years	13
First World War	14
Inter-war years	16
Second World War	18

22	The story of *Alliance* and the A-class submarines
A-class design and construction	24
A-class submarines in service	26
Alliance in service	30
Nuclear dawn and the demise of diesel-electric submarines	41

44	Anatomy of *Alliance*
How a submarine works	48
Construction	51
General internal arrangements	55
External tanks	62
Tank blowing and venting systems	66
High-pressure air system	69
Main telemotor (hydraulic) system	71
Auxiliary machinery and equipment	73

82	*Alliance*'s propulsion
Diesel main engines	84
Main electric motors and associated components	92
Snort induction and exhaust systems	98
Diesel fuel oil and compensating systems	101
Battery and electrical distribution	103

106	*Alliance*'s weapons
Torpedoes	108
Mk XXIII quick-firing 4in gun	114
Oerlikon 20mm anti-aircraft gun	118
Additional armament and ammunition	119

120	*Alliance*'s operational equipment
Steering gear and control systems	122
Hydroplane gear	124
Periscopes	125
Radar systems	131
ASDIC or sonar systems	132
Capstans and anchor gear	132

134	Operating *Alliance*
Diving the boat	136
Surfacing the boat	137
Running the boat when submerged	138
Running silent, running deep	139
Bringing the boat to periscope depth	139
Snorting	140
Running the boat when surfaced	142
Torpedo attack	142
Diving stations	144

146	Manning *Alliance*
The crew and their duties	148
Living in the boat	152

158	Submarine escape
Submarine loss	160
'Sub-miss' and 'Sub-sunk'	161
Submarine escape training tower (SETT)	161
Buoyant exhaling ascent	162
Bringing attention to a stricken submarine	163
Escape procedure	164

166	Restoring *Alliance*
External repair and restoration	170
Internal conservation	175

178	Appendices
1 – List of sources and further reading	178
2 – Glossary of terms	180
3 – A-class submarine specifications	181
4 – A-class submarine weights and capacities	182
5 – Preserved submarines open to visitors in the UK	183
6 – Preserved submarines open to visitors worldwide	184

187	Index

OPPOSITE This dramatic bow-on view of the newly restored HMS *Alliance* was taken at the Royal Navy Submarine Museum, Gosport, in 2014. *(Jonathan Falconer)*

Introduction

Newly restored and the prize exhibit at the Royal Navy Submarine Museum, HMS *Alliance* has a unique place in history as the only surviving British Second World War patrol submarine. She is a fitting memorial to 5,300 British submariners who gave their lives in service.

HMS *Alliance* is the third vessel to bear this name in the Royal Navy. The first, originally named *Alliante,* was a 20-gun ship captured from the Dutch off the coast of Norway on 22 August 1795; renamed *Alliance*, she served in the Royal Navy as a stores ship until sold in May 1802. The second *Alliance*, the lead vessel of a class of four Admiralty tugs built in 1910, was scuttled in Hong Kong in 1941 to prevent her capture by the invading Japanese forces.

Built for service in the Second World War, the A-class submarine *Alliance* is the only surviving member of her type. Today, at her home at the Royal Navy Submarine Museum, she stands testament to the courageous men who fought in similar 'boats' during the Second World War for defending the freedom we enjoy today.

The rationale for choosing *Alliance* as the subject for a Haynes Manual is based on four factors:

- *Alliance* epitomises the peak of British submarine development based on the operational and technical experiences attained through years of combat during the Second World War.
- It precedes the post-war demise of the conventional diesel-electric 'submersible boats' and the ascendancy of nuclear-powered submarines, which are capable of sustained submerged operation.
- *Alliance* is the first British submarine to have been conserved and forms the nucleus of the collections maintained for public display at the Royal Navy Submarine Museum at Gosport, Hampshire.
- This Haynes Manual will generate awareness of the continuous maintenance and restoration needed to preserve *Alliance* as a lasting tribute to the unique breed of men who have served, and continue to serve, in the Royal Navy Submarine Service.

It is worth recounting briefly how *Alliance* came to be at the Royal Navy Submarine Museum. Following her decommissioning from active service (see Chapter 2), in March 1973 she replaced HMS *Tabard* as the static display submarine and floating classroom alongside the jetty at HMS *Dolphin*, the shore establishment sited at Fort Blockhouse, Gosport. *Dolphin* was the home of the Royal Navy Submarine

OPPOSITE HMS *Alliance* – built at the end of the Second World War and restored in the 21st century as a lasting tribute to British submariners. *(All photos RN Submarine Museum (RNSM) unless credited otherwise)*

BELOW *Alliance* under scaffolding at the start of her £7 million restoration. *(Jonathan Falconer)*

ABOVE **Contractors at work on the forward hydroplanes.**
(Jonathan Falconer)

Service from 1904 to 1999 and the Royal Navy Submarine School was based there until its relocation to HMS *Raleigh* at Torpoint, Cornwall.

Following a Government White Paper proposal to set up a permanent submarine museum at *Dolphin*, in February 1978 *Alliance* was transferred to the new Royal Navy Submarine Museum to serve as a permanent memorial to the officers and men of the Royal Navy Submarine Service who had lost their lives in wartime and peacetime.

Before going on display to the public, *Alliance* required some repairs and restoration. In August 1979 she was towed to Vosper Ship Repairs Ltd at Southampton to prepare her hull for permanent display, returning to the museum later that year. Work then started to mount *Alliance* out of the water for greater accessibility, and internal restoration work was carried out to bring her back to active-service condition.

Preserving *Alliance* for future generations

In recent years her condition had become a cause for concern and plans were put in place for a new – and much more extensive – phase of restoration work together with improvements to the visitor experience in terms of display and interpretation. The total cost would be £7 million and the breakthrough in a three-year fundraising campaign came on 30 May 2011 when, following formal applications submitted to the Heritage Lottery Fund, it was announced that *Alliance* would receive a £3.4 million grant for repair to her bow and stern, and work to address extensive surface corrosion to her hull and casing (see Chapter 10). All external restoration work was completed by July 2013, and the accompanying interior conservation programme was finished in time for *Alliance* to reopen to the public in April 2014.

Since the Royal Navy Submarine Museum opened in 1982, *Alliance* has always been one of the most accessible historic submarines in the world, and now today's visitors are given the best opportunity to fully understand and appreciate her. Although visitor surveys over the years had always confirmed that guided tours by ex-submariners are well appreciated, the following improvements have occurred:

- There is now public access to more parts of the submarine.
- To help visitors understand life on board, the living quarters, conning tower and casing have been made accessible.
- New interpretation, with state-of-the-art lighting and sound, bring the submarine to life.

ABOVE HMS *Alliance*, restored to her former glory and open once more to the public.

A visit to *Alliance* begins with a brand-new film with voiceover by British actor Ian McShane highlighting life on board from the immediate aftermath of the Second World War through the Cold War into the 1970s. Visitors then take a chronological journey through every decade of the submarine's service from the 1940s to the 1970s. They are guided through the submarine by one of a team of veteran submariners who tell their stories of working beneath the waves. Sounds and smells capture some of atmosphere inside a working submarine, and visitors can use the working periscope to view Portsmouth harbour.

Chris Munns, Director at the Royal Navy Submarine Museum, said, 'A visit onboard HMS *Alliance* will assault all the senses and really bring to life what it is like to work and live on a submarine. We are very proud of HMS *Alliance* and delighted that she has been saved for future generations.'

The Royal Navy Submarine Museum

The humble beginnings of the Royal Navy Submarine Museum were born from the gatherings of a Submarine Branch Collection contained above St Ambrose Church in HMS *Dolphin*. Although formally opened in 1963, security necessities unfortunately restricted public access for many years. During this period the museum well served those – including the author – under training for the submarine service, helping to instill in them the strong sense of camaraderie unique to submariners.

The museum was officially recognised by the Ministry of Defence in 1967, its first full-time curator was appointed in 1968, and it was formally registered as a charity in 1970. The museum was moved outside the confines of *Dolphin* in 1978, allowing full public access. It was at this time that *Alliance*, the Royal Navy's training and static display submarine, was donated to the museum.

The complex that is now recognised as the Royal Navy Submarine Museum opened in August 1981 with *Alliance* as its principal exhibit. Since then, the acquisition of more submarines and supporting collections has extensively expanded the collection (see Appendix 5). Two highlights have been the restoration of a Second World War X-class midget submarine and the discovery and salvage of the remains of the Royal Navy's first submarine, *Holland 1*, built in 1901, which is now conserved and fully accessible to the public within its own climate-controlled building.

Chapter One

British submarine development, 1901–45

The Royal Navy was marginally slower than other naval powers in adopting submarines, but the birth in 1901 of the Royal Navy Submarine Service – initially equipped with just five 'Holland' boats built to an American design – quickly led to a long line of British-built submarines, mostly of diesel-electric propulsion.

OPPOSITE T-class boat HMS *Thunderbolt* in harbour.

At the end of the 19th century the Royal Navy remained the predominant naval power in the world. By this time other naval powers – Russia, France and Germany in particular – were beginning to develop submarines, but initially there was reluctance at the Admiralty to do the same, Admiral Sir Arthur Wilson allegedly having described submarine warfare as 'unnecessary, ungentlemanly, piratical and damned un-English'.

At the forefront in overturning this reluctance was Admiral Sir John Arbuthnot 'Jacky' Fisher. Farsighted in leadership and an innovative reformer, Fisher firmly believed that if the Royal Navy was to maintain its supremacy over other rising world powers, it had to adopt new methods of warfare and accept the fact that it needed to create its own fleet of submarines. In addition, Fisher also oversaw the introduction of a revolutionary class of 'Dreadnought' battleships. By forcibly driving through these two separate initiatives, Fisher completely reinvented the Royal Navy to meet the demands of the early 20th century.

Birth of the Royal Navy Submarine Service

At the end of the 19th century France was the only nation with the right kind of technical know-how to develop a submarine fleet that could pose a threat to Britain. To counteract this, the Royal Navy needed to build up a new fleet of submarines, and quickly. Although the best solution was to purchase boats direct from the Americans to a design by John P. Holland, the Admiralty had concerns about the political difficulties of sourcing from the USA at a time when Britain did not yet fully trust the ascendancy of American naval power.

Holland was at the forefront of submarine development, having pioneered the use of the internal combustion engine (rather than steam) in his first submarine, *Fenian Ram* of 1881, and then gone on to refine his ideas to the point where *Holland VI* of 1897 had a battery-powered electric motor for submersible operation in addition to an internal combustion engine for surface running, and it also featured a conning tower for observation while submerged.

In December 1900 the Royal Navy signed a contract with the American Electric Boat Company for five 'Holland-design boats' to be built in Britain and the orders for construction were given to Vickers, Sons & Maxim of Barrow-in-Furness. Ever since then, incidentally, Vickers (now BAE Systems) has remained the leading British shipbuilder in the field of submarine construction for the Royal Navy, with *Alliance* herself having been built by the company.

The keel of *Holland 1* was laid down in February 1901. To maintain secrecy the boat was constructed in a building named 'Yacht Shed', and components fabricated in the general yard were covertly marked for 'Pontoon No 1'. When *Holland 1* was commissioned, shortly after launch on 2 October 1901, the Royal Navy Submarine Service was born under the authority of the newly appointed Inspecting Captain of Submarine Boats, the experienced torpedo officer Captain Reginald Bacon. *Holland 1* dived for the first time in an enclosed basin in March 1902 and began sea trials in April with initial training being supervised by American Frank Cable and his team.

In September 1902 *Holland 1* and other completed 'Holland' boats arrived at Portsmouth together with the Dryad-class torpedo gunboat HMS *Hazard* acting as their depot ship. Collectively these vessels formed the 'First Submarine Flotilla' under Captain Bacon's command. Knowing how dangerous the new submarines could be, Bacon diligently trained his small band of volunteer officers and men. Despite initial accidents and

BELOW *Holland 1.*

disappointments, within a few months Captain Bacon reported, 'Even these Little Boats would be a terror to any ship attempting to remain or pass near a harbour holding them.'

Holland 1 was decommissioned and sold semi-stripped of fittings in 1913 to T.W. Ward Ltd for £410, but was lost while being towed to the scrapyard when she foundered in bad weather off the Eddystone lighthouse. Located on the seabed in 1981 and raised in November 1982, she has been conserved for display in the Royal Navy Submarine Museum.

The first ten years

The five Holland boats did need improvements in service and two stand out. Firstly, Captain Bacon enhanced control by separating rudder and hydroplane operation, with two operators rather than one. Secondly, viewing by means of a 'fixed' optical tube – a forerunner of the periscope – superseded glass scuttles.

In 1902 the first British-designed A-class submarines were introduced by Vickers, with HMS *A1* launched at Barrow-in-Furness in July 1902. Colloquially known as 'Fisher's toys', these A-class boats were propelled by a battery-powered electric motor when submerged (like the Holland boats) and a Wolseley petrol engine when surfaced, though the last of the class, *A13*, launched in April 1905, was fitted experimentally with a Vickers diesel engine; as well as being far more robust and reliable, diesel engines (and their lower-octane fuel) presented much less fire risk than their soon-to-be-obsolete petrol counterparts.

Vickers followed the A-class with 11 rather similar B-class submarines built in the period 1904–06. Then came the C-class, which marked the end of development of the Holland-type petrol-electric design. The C-class was very much larger than its predecessor, with length increased from 63ft 10in (19.46m) to 143ft 2in (43.64m) and beam from 11ft 9in (3.58m) to 13ft 6in (4.11m). With a crew of 16 men rather than 8, endurance was more limited, and in addition the C-class was a poor vessel when surface running because it had only 10 per cent reserve of buoyancy above its surface displacement, but the spindle-shaped hull provided good underwater performance. In total 38 C-class submarines were built and many served in the First World War in the Baltic.

The next development was the technically more advanced D-class, of which ten were built and completed in the period 1907–10. This was the first class of Royal Navy submarines driven by diesel engines, which proved more robust and reliable than petrol engines. Each of the two propulsion units comprised a 550bhp electric motor and a 1,750bhp diesel engine. Unlike its predecessors, the D-class was designed to operate significant distances beyond coastal waters and had a potential range of 2,500 nautical miles when running on the surface. These were also the first British submarines to have wireless transmitters; the aerial, which was attached to the conning tower mast, was

BELOW HMS *D8* under way. Introduced in 1910, the D-class were the first British boats to be driven by robust diesel engines, improving reliability, range and endurance. They were also the first boats to have wireless telegraphy.

ABOVE HMS *K6*. Although designed as fleet submarines, engineering technology of the time was not sufficiently advanced to guarantee the success of the K-class.

lowered before diving. Another innovation, introduced from HMS *D6*, was the addition of a deck gun, a quick-firing 76mm 12-pounder, and other armament comprised three 18in torpedo tubes (two bow, one stern).

First World War

Fifty-six examples of the E-class, an improved version of the D-class, were built in the period 1912–16 and formed the backbone of the Royal Navy submarine fleet throughout the First World War, successfully operating in the North Sea, the Baltic and the Sea of Marmara during the Gallipoli campaign. These boats also served in the newly created Royal Australian Navy. E-class boats were more heavily armed than the D-class, early examples ('Group 1') having four 18in torpedo tubes (one bow, two beam, one stern) and later examples ('Group 2' and 'Group 3') having five torpedo tubes (two bow, two beam, one stern).

Next came the innovative steam-driven K-class submarines, which proved to be controversial. Initially designed in 1913, these were 'fleet submarines' that were intended to operate with the surface fleet by sweeping ahead of squadrons engaged in a fleet action and then rapidly manoeuvring behind the enemy ships to ambush them as they retreated. Needing a surface speed of 24kts to keep pace with the fleet, they required a steam propulsion plant comparable with the motive power of the surface ships they accompanied.

The K-class submarines were fitted with twin 10,500bhp oil-fired Yarrow boilers, each powering a Brown-Curtis or Parsons geared steam turbines driving two shafts. Steam power necessitated having retractable funnels that closed down before diving, after which the boat proceeded, like a conventional submarine, at a maximum speed of 8kts on electric power, by means of four 1,440bhp motors. There was also an 800bhp Vickers diesel generator to charge the batteries and provide emergency motive power in the event of failure of the steam propulsion.

Mammoth in comparison to their contemporaries, the K-class submarines were 338ft (103m) long and displaced 1,980 tons on the surface and 2,566 tons submerged. The design featured a double hull, providing a reserve buoyancy of 32½ per cent. However, there were design flaws associated with the great size of these submarines.

Control and 'depth-keeping' proved difficult, mainly because efficient hydraulic control systems had not yet been fully developed. Because the K-class's designed maximum diving depth of only 200ft (60m) was much less than the 338ft (103m) hull length, problems occurred in relation to angle of dive. A 10-degree angle would cause a difference in depth between bow and stern of around 60ft (18m), but a 30-degree angle would produce a 170ft (51m) difference – so that the bow would be close to maximum safe depth while the stern would still be near the surface. Even worse, the eight internal bulkheads were only designed and tested to stand a pressure of 35psi, which is equivalent to a depth of only 70ft (20m).

ABOVE HMS *M2* with her Parnall Peto seaplane taking off. The M-class were designed as monitors to work with the fleet. *M2* was fitted with a hangar enclosing a Parnall Peto seaplane that was used for reconnaissance work. Her counterparts *M1* had a 12in gun with which to attack large warships, while *M3* was configured as a minelayer.

There were other inherent problems. Surface running at high speed tended to drive the bow too deep into the water, worsening the already poor sea-keeping capabilities of these boats; this deficiency was alleviated by fitting a bulbous 'swan bow'. A regular operational problem in heavy seas was ingress of water through the short twin funnels putting out the boiler fires. The high temperatures generated within the boiler rooms necessitated larger fans to be installed. Initial designs also incorporated a pair of swivel-mounted 18in (457mm) torpedo tubes fitted on the casing for surface use at night. These were prone to damage in rough seas and were later removed.

Involved in many accidents, the K-class submarines acquired notoriety that earned them the nickname of 'Kalamity' class. Though none of the 18 built was lost through enemy action, 6 sank in accidents. Only one had the opportunity to engage an enemy vessel, firing a torpedo and hitting a U-boat amidships, but unfortunately the torpedo failed to explode.

Although historically maligned as seagoing follies, from a submarine engineer's perspective I certainly believe that despite their mechanical shortcomings, the K-class submarines were an ingenious and progressive experiment in maximising power and armament into a formidable submersible machine. Their only failing was that technology had not advanced sufficiently enough in hydraulics and electro-solenoid valve operation for these boats to fulfil their potential capability with a greater margin of safe operation and control. It was to be another 50 years before the concept of 'fleet submarines' was fully realised, when, in 1960, the nuclear-powered hunter-killer fleet submarine HMS *Dreadnought* entered service with the Royal Navy.

The L-class was an improved version of the E-class and 43 were built in three phases between 1917 and 1919; some were later converted to act as minelayers. These were the first submarines to be fitted with the new, larger 21in (533mm) torpedoes. All L-class vessels had a quick-firing 4in (102mm) deck gun and some later examples had two.

Next developed during the First World War was the diesel-electric M-class 'monitor', a term used for a relatively small vessel carrying a disproportionately large gun – in this case a 12in (305mm) Mk IX mounted in a turret forward of the conning tower. The gun could be fired from periscope depth but the submarine had to resurface for it to be reloaded. In practice the concept was not particularly effective and only three M-class submarines were built in the period 1917–18.

There was also the one-off *Nautilus*, which was the largest submarine built for the Royal Navy at the time and the first to be given a name. Weighing 1,000 tons and capable of 20kts (37kph), *Nautilus* (later renamed *N1*) had a large double hull instead of the saddle tanks that had been used for previous designs. Although her keel was laid down in 1913, *Nautilus* was not completed until 1917 and she was never used for active service, merely serving as a depot ship based at Portsmouth.

The final development during the First World War was the R-class, which was introduced in 1917 and of which 12 were built. These were the first hunter-killer subs and, although small, these boats were specifically designed with a battery capacity and optimised streamlined hull shape enabling underwater performance to attack and sink enemy submarines. Single-shafted, driven by an eight-cylinder 480hp diesel and two 1,200shp electric motors, the R-class had an additional single motor for low-speed running. These boats had a surface speed of 9½kts (17.6 km/h) and a submerged speed of 14kts (26km/h). With a submerged endurance of one hour at 14kts (26 km/h) they were armed with 6 x 18in bow torpedo tubes and carried 12 reload torpedoes. Crewed by just 2 officers and 20 ratings, these craft carried new operational innovations in the form of sensors and processing systems comprising bow hydrophone (sonar) arrays.

Inter-war years

Few important technological advances were made in submarine propulsion or weaponry during the years between the wars, but one exception was HMS *X1*, laid down at Chatham in 1921 and completed in 1925. The world's largest submarine at the time, *X1* was designed as a 'commerce raider' – a vessel for attacking merchant shipping convoys and their escorts – and was based on the uncompleted German U-173 class of 2,000-ton 'U-Cruisers'. In 1922, however, the Washington Naval Treaty, to which Britain was a signatory, banned attacks on merchant shipping as a result of outrage over German U-boat actions during the war, so there was considerable sensitivity and secrecy surrounding *X1*'s development. In fact *X1*, for all her formidable weapon systems, suffered persistent engine problems in service and was eventually scrapped in 1936.

Three new classes of relatively conventional submarines were introduced during the late 1920s and early 1930s, all of very similar design and with the improved endurance capability needed for their intended service in the Far East: these were the Odin-class (1928, nine built), the Parthian-class (1929, six built) and the Rainbow-class (1930, four built).

Designed to patrol the North Sea and Mediterranean, the S-class was a smaller submarine that became the Royal Navy's most numerous single class of submarine, with 62 constructed over a 15-year period. The first, HMS *Swordfish*, was laid down at Chatham on 1 December 1930, launched on 10 November 1932 and commissioned just 18 days later; a total of 12 S-class boats were built by 1935. It was the approach of war that brought further construction in large numbers, with 50 boats

BELOW HMS *X1* was designed as a submersible commerce raider. She was armed with four QF 5.2in Mk I guns mounted in twin unarmoured turrets and six 21in bow torpedo tubes. When launched she was the largest submarine in the world.

to a slightly larger and more heavily armed specification constructed in the period 1939–43. Armament comprised six 21in internal bow torpedo tubes with 12 reload torpedoes, one 3in deck gun and one .303-calibre machine gun.

Many S-class submarines served with distinction during the Second World War. The most famous, HMS *Seraph*, operated with the 8th Submarine Flotilla in the Mediterranean and, after being appointed to special duties, was assigned to operations codenamed 'Torch', 'Flagpole' and 'Mincemeat'. 'Flagpole' involved delivering a group of senior US officers, led by Lieutenant General Mark W. Clark (General Dwight D. Eisenhower's deputy), to the Algerian coast in October 1942 for secret negotiations with Vichy French officers about the invasion of North Africa, and then safely collecting them again. 'Mincemeat' was *Seraph*'s part in 'Barclay', the plan designed to mislead the Germans that the Allies intended to land in Greece and Sardinia rather than Sicily. *Seraph* carried a corpse with a briefcase containing false 'secret documents' and discharged this off the Spanish coast; the Spanish authorities passed the fake information to the Germans, who did indeed divert defending forces from Sicily as a consequence.

Besides the S-class boats, the 1930s saw the introduction of three River-class submarines (also known as Thames-class) built as a last attempt by the Admiralty to produce 'fleet submarines' – submarines fast enough to operate as part of a fleet and thus requiring good surface speed. The design compromised on diving depth to keep weight down and two supercharged Ricardo diesel engines produced maximum output of 10,000bhp, good enough for 22kts (41kph).

Also built during the 1930s were six Grampus-class minelaying submarines. These boats are sometimes known as Porpoise-class, named after the prototype, which was built in 1932; the remaining boats were built much later, in the period 1936–38.

The T-class boats were a very successful design, with the 53 examples produced playing a major role in the Royal Navy's submarine operations. The first, named *Triton*, was laid down on 28 August 1936 by Vickers-Armstrong at Barrow-in-Furness and launched on 5 October 1937, and construction of further submarines – sometimes called Triton-class submarines – continued during the war. These submarines had ten bow torpedo tubes (six internal, four external), enabling them to fire a larger salvo of torpedoes than any previous operational submarine, the reason being that enemy anti-submarine measures would mean that attacks on surface warships would have to be carried out at longer range and therefore with less accuracy.

BELOW S-class, HMS *Solent*. This small class of submarine was designed for patrolling the waters of the North Sea and Mediterranean, although *Solent* spent much of her career in the Far East during the Second World War. *(Jonathan Falconer collection)*

LEFT T-class, HMS *Thorn* **in harbour. Fifty-three Triton-class boats were built before and during the Second World War. T-boats had a highly successful operational career and they heavily influenced the design of** *Alliance* **and her class.**

The last T-class boat in Royal Navy service, albeit non-operational, was HMS *Tabard* (laid down in 1943), which was permanently moored as a static training submarine at HMS *Dolphin*, Gosport, from 1969 to 1974, when she was replaced by HMS *Alliance*.

Unfortunately, despite their reliability, three T-class boats were accidentally lost, with grave consequences (see Chapter 9). The first, *Thetis*, sank with the loss of 99 lives on 1 June 1939 while undertaking acceptance diving trials in Liverpool Bay; salvaged, repaired and commissioned under the new name of *Thunderbolt*, she served with distinction in the Atlantic and Mediterranean until she was lost in action with all hands on 14 March 1943 – making her one of very few military vessels in service history to be lost twice with her entire crew. The second, *Truculent*, was sunk in the Thames estuary on 12 January 1950 with the loss of 79 lives after being accidentally rammed at night by the Swedish oil tanker *Divina*. The third, *Totem*, which was sold to the Israeli Navy in 1965 and renamed INS *Dakar,* undertook sea and diving trials off Scotland before sailing to Israel on 9 January 1968 but disappeared with all hands off Crete.

Second World War

Additional to the operationally successful S-class and T-class boats, two other classes – U and V – were introduced during the Second World War to supplement the Royal Navy's hard-pressed submarine fleet.

The U-class boats were small, weighing only 630 tons, and were originally intended to be unarmed training vessels to replace the ageing H-class boats. The first three examples – sometimes called the Undine-class after the first submarine built – were ordered in 1936,

LEFT HMS *Thunderbolt* **and her crew in 1942.**

ABOVE U-class, HMS *Upstart*. Originally designed as training vessels, 49 of these short-hull U-class submarines were built during the War Emergency 1940 and 1941 programmes. As the only vessel of the Royal Navy bearing this name she was loaned to the Greek Navy after the war and renamed *Amphitriti*. (Jonathan Falconer collection)

but during construction, with the threat of war growing, they were modified to accommodate four internal and two external bow torpedo tubes. Just before war started, another 12 U-class 'group two' boats were ordered, eight of them lacking the external torpedo tubes because these were found to create an excessive bow wave that made depth-keeping too difficult at periscope depth. A further 34 'group three' vessels were built under the 'War Emergency 1940 and 1941 Programmes, Short Hull', making a final total of 49 U-class boats; vessels of groups two and three were sometimes called the Umpire-class after the first group two submarine built.

Most of the U-class boats served in the 10th Submarine Flotilla based at Malta and proved to be useful warships, though losses were high at 19 of the 49 built. One of the losses, in April 1942, was the renowned HMS *Upholder*, which in 16 months carried out 24 patrols and sank 119,000 tons of enemy shipping, including three U-boats, a

LEFT HMS *United* (left) and *Unison* (centre), with the bows of *Unseen* in the foreground, of the Royal Navy's 10th Submarine Flotilla at Malta in January 1943. The most famous boat of this squadron was HMS *Upholder*, commanded by Lieutenant Commander Malcolm David Wanklyn VC, DSO & Two Bars, which became the most successful British submarine of the Second World War. (IWM A14512)

HMS *Ultimatum* at harbour stations. The twin-shafted U-class was driven by two Paxman Ricardo diesel generators and electric motors providing a surface speed of 11¼kts and a submerged speed of 10kts. Armed with four 21in torpedo bow tubes they carried 8 to 10 reload torpedoes and 1 3in gun. The average crew complement was just 26 men.

destroyer and 15 transport ships. Throughout her service *Upholder* was in the command of Lieutenant Commander David Malcolm Wanklyn, who received a Victoria Cross for attacking a well-defended convoy on 25 May 1941 and sinking the Italian liner *Conte Rosso*.

V-class submarines, built under the 'U-Class Long Hull 1941–42 Programme', followed directly on from the U-class programme and their design was very similar. The main change was to the hull, with a lengthened stern and finer line to the bow giving improved underwater handling and reduced noise from water turbulence; in addition the pressure hull plating was ¾in (19.1mm) rather than ½in (12.7mm) to allow deeper diving. Although Vickers-Armstrong received orders to construct 40 V-class boats, only 22 were completed.

And so this story arrives at the A-class submarines, which are covered in the next chapter.

CLASSES OF BRITISH SUBMARINE, 1901–45

Class	Date built	Number built
Petrol-electric		
Holland-class	1901–02	5
A-class	1902–05	13
B-class	1904–06	11
C-class	1906–10	38
Diesel-electric		
D-class	1908–12	8
E-class	1912–16	58
F-class	1913–17	3
S-class	1914–15	3
V-class	1914–15	4
W-class	1914–15	4
G-class	1915–17	14
H-class	1915–19	44
J-class	1915–17	7
L-class	1917–19	34
M-class	1917–18	3
Nautilus-class	1917	1
R-class	1918	12
HMS *X1*	1921	1
Odin-class	1926–29	9 (subclasses: Oberon 1, Oxley 2, Odin 6)
Parthian-class	1929	6
Rainbow-class	1930	4
S-class	1931–45	62 (subclasses: Swordfish 4, Shark 8, Seraph 33, Subtle 17)
Thames-class	1932	3
Grampus-class	1932–38	6
T-class	1937–45	52 (subclasses: Triton 15, Tempest 15, Taciturn 22)
Undine (U)-class	1937–38	3
P611-class	1940	4
Umpire (U)-class	1940–43	37
Vampire (V)-class	1943–44	22
Amphion (A)-class	1945–47	16
Steam-electric		
Swordfish	1916–22	1
K-class	1916–19	22
Midget		
X-class	1943–44	20
XE-class	1944	6

Chapter Two

The story of *Alliance* and the A-class submarines

HMS *Alliance* was designed during the Second World War for service in the Far East and was launched in July 1945, just as victory was achieved. She then began a distinguished 28-year career during the Cold War until she retired and became the centrepiece of the Royal Navy Submarine Museum.

OPPOSITE HMS *Alliance* leaves HMS *Dolphin* for the HMS *Affray* Memorial Service. Note the boys' training ship TS *Foudroyant* in the harbour, in which the author once spent a week as a boy seaman (*Foudroyant* has since been restored to her original form as the 1817 frigate HMS *Trincomalee*).

ABOVE *Alliance* leaving Portsmouth in 1953.

In June 1941 Prime Minister Winston Churchill requested a new submarine programme that would require the rapid construction and delivery of a large number of submarines to meet the evolving needs of the war. Following the Japanese attack on Pearl Harbor on 7 December 1941, Britain had to protect her interests in the Far East and Pacific and to do this the Royal Navy needed submarines capable of operations against Japan in a very extensive new theatre of war that encompassed Australia, Burma, Malaya, Singapore and Hong Kong.

A-class design and construction

Planned as part of the '1943 Emergency Building Programme', the Amphion-class – or A-class – diesel-electric submarines were designed and equipped for operations against Japan in the vast Pacific region. Capable of a high surface speed, they had the capacity and endurance to undertake long-range patrols, and they could dive deeper than the previous S-class and T-class submarines.

The maximum speed of an A-class submarine when surfaced was 18½kts and the range of 10,500 nautical miles (at 11kts) was outstanding. Operating depth was 500ft (c.150m) but a maximum deep diving depth of 750ft (c.230m) was permissible; maximum speed when submerged was 8 knots. The A-class submarines were powered by two Vickers eight-cylinder diesel engines delivering 2,150bhp each, and for normal submerged running there were two English Electric motors producing 625bhp each.

Non-Vickers-built A-class boats were fitted with two alternative engine types with slightly differing weights, specifications and running criteria. Those built by Cammell Laird or Scotts (*Acheron*, *Aeneas*, *Affray*, *Alaric*, *Artemis* and *Artful*) were fitted with Admiralty-designed 'straight 8's'. Those built in Devonport Dockyard (*Ace* and *Achate*) were fitted with engines designed by John Brown. From my experience, the Admiralty engines restricted walking space along the engine room centreline.

Initially armed with 10 torpedo tubes (with 20 reload torpedoes), a 4in deck gun and a 20mm Oerlikon cannon, the A-class vessels were among the most formidably armed submarines of the time and were well capable of meeting potential combat requirements.

Climatic conditions in the Far East necessitated the provision of air conditioning in order to provide some relative comfort for the crew, and all accommodation was placed as far as possible from the engine room. The crew of an A-class submarine normally numbered 61, 6 of whom were officers.

The initial build programme was for 46 A-class boats, as listed in the accompanying table. However, only 16 were commissioned into service with the Royal Navy, 28 boats being cancelled before build started and 2 remaining unfinished. Construction was carried out at five shipyards: Vickers-Armstrong at

A-CLASS SUBMARINE BUILD PROGRAMME

No	Name	Pennant no	Builder/yard	Keel laid	Launched	Completed
1	Acheron	P 411	HM Dockyard, Chatham	26.8.44	25.3.47	17.4.48
2	Adept	P 412	HM Dockyard, Chatham	Cancelled	–	–
3	Ace	P 414	HM Dockyard, Devonport	–	–	Uncompleted
4	Alcide	P 415	Vickers-Armstrong, Barrow	2.1.45	12.4.45	18.10.46
5	Alderney	P 416	Vickers-Armstrong, Barrow	6.2.45	25.6.45	10.12.45
6	Alliance	P 417	Vickers-Armstrong, Barrow	13.3.45	28.7.45	14.5.47
7	Ambush	P 418	Vickers-Armstrong, Barrow	17.5.45	24.9.45	22.7.47
8	Auriga	P 419	Vickers-Armstrong, Barrow	7.6.44	29.3.45	12.1.46
9	Affray	P 421	Cammell Laird, Birkenhead	16.1.44	20.4.45	12.11.45
10	Anchorite	P 422	Vickers-Armstrong, Barrow	19.7.45	22.1.46	18.11.47
11	Andrew	P 423	Vickers-Armstrong, Barrow	13.8.45	6.4.46	16.3.48
12	Andromanche	P 424	Vickers-Armstrong, Barrow	Cancelled	–	–
13	Answer	P 425	Vickers-Armstrong, Barrow	Cancelled	–	–
14	Aurochs	P 426	Vickers-Armstrong, Barrow	21.6.44	28.7.45	7.2.47
15	Aeneas	P 427	Cammell Laird, Birkenhead	10.10.44	25.10.45	31.7.46
16	Antagonist	P 428	Vickers-Armstrong, Barrow	Cancelled	–	–
17	Antaeus	P 429	Vickers-Armstrong, Barrow	Cancelled	–	–
18	Anzac	P 431	Vickers-Armstrong, Barrow	Cancelled	–	–
19	Aphrodite	P 432	Vickers-Armstrong, Barrow	Cancelled	–	–
20	Achates	P 433	HM Dockyard, Devonport	–	–	Uncompleted
21	Admirable	P 434	Vickers-Armstrong, Tyneside	Cancelled	–	–
22	Approach	P 435	Vickers-Armstrong, Barrow	Cancelled	–	–
23	Arcadian	P 436	Vickers-Armstrong, Barrow	Cancelled	–	–
24	Ardent	P 437	Vickers-Armstrong, Barrow	Cancelled	–	–
25	Argosy	P 438	Vickers-Armstrong, Barrow	Cancelled	–	–
26	Amphion	P 439	Vickers-Armstrong, Barrow	14.11.43	31.8.44	27.3.45
27	Alaric	P 441	Cammell Laird, Birkenhead	31.5.45	18.2.46	11.12.46
28	Atlantis	P 442	Vickers-Armstrong, Barrow	Cancelled	–	–
29	Agile	P 443	Cammell Laird, Birkenhead	Cancelled	–	–
30	Asperity	P 444	Vickers-Armstrong, Tyneside	Cancelled	–	–
31	Austere	P 445	Vickers-Armstrong, Tyneside	Cancelled	–	–
32	Aggressor	P 446	Cammell Laird, Birkenhead	Cancelled	–	–
33	Agate	P 448	Cammell Laird, Birkenhead	Cancelled	–	–
34	Astute	P 447	Vickers-Armstrong, Barrow	4.4.44	30.1.45	30.6.45
35	Artemis	P 449	Scotts, Greenock	28.2.44	26.8.46	15.8.47
36	Abelard	P 451	HM Dockyard, Portsmouth	Cancelled	–	–
37	Acasta	P 452	HM Dockyard, Portsmouth	Cancelled	–	–
38	Alcestis	P 453	Cammell Laird, Birkenhead	Cancelled	–	–
39	Aladdin	P 454	Cammell Laird, Birkenhead	Cancelled	–	–
40	Aztec	P 455	Vickers-Armstrong, Tyneside	Cancelled	–	–
41	Artful	P 456	Scotts, Greenock	8.6.44	22.5.47	23.2.48
42	Adversary	P 457	Vickers-Armstrong, Tyneside	Cancelled	–	–
43	Asgard	P 458	Scotts, Greenock	Cancelled	–	–
44	Awake	P 459	Vickers-Armstrong, Tyneside	Cancelled	–	–
45	Astarte	P 461	Scotts, Greenock	Cancelled	–	–
46	Assurance	P 462	Scotts, Greenock	Cancelled	–	–

LEFT *Alliance* and other boats alongside the depot ship. Note the stern torpedo tubes.

Barrow-in-Furness (ten boats), Cammell Laird at Birkenhead (three boats), Scotts Shipbuilding and Engineering Company at Greenock (two boats), HM Dockyard at Chatham (one boat) and HM Dockyard at Devonport (the two uncompleted boats).

Unlike the riveted plate-on-frame construction previously used, the A-class submarines were designed for rapid construction, in keeping with the needs of war, and featured an entirely welded hull that could be fabricated in sections, a technique that had been used for German U-boats but was new for a British design, albeit trialled on the latter-built T-class boats. From laying the keel to launch took only 8 months, compared with around 15 months for the earlier T-class boats. Welding quality had been improved by advances in X-ray technology.

A-class submarines in service

Despite being designed for operations against Japan in the Pacific, the war ended before any A-class submarines saw active service. In fact only two A-class boats were completed before the end of the war: *Amphion* was launched in August 1944 and *Astute* in January 1945.

Development of the snort mast

When the A-class boats were being developed, the Royal Navy used the opportunity to experiment with a snort mast. Initially conceived by the Dutch and further developed by the Germans for their U-boats, a snort mast permitted a submarine to run submerged under diesel power (from its main engines) by drawing in air through a snorkel. Seven A-class boats, including *Alliance*, were built with a snort mast, and the nine other boats had a snort mast fitted by 1949. The operational advantages of running

LEFT Close-up of HMS *Andrew* midships showing snort mast (foreground), attack periscope (left) and search periscope (middle).

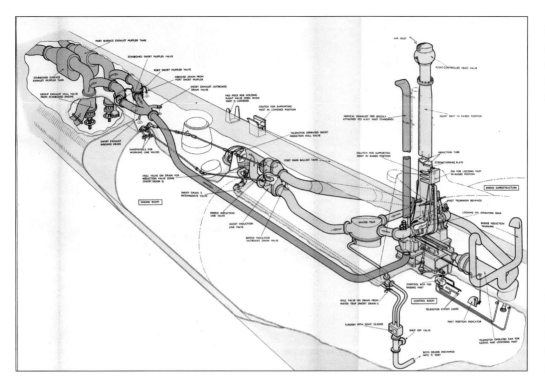

LEFT General arrangement drawing of the snort system in the A-class. *(Plate 4, BR.1963/8)*

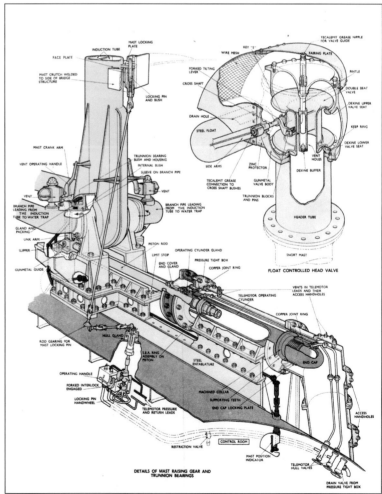

BELOW General arrangement of the snort mast and its raising gear. *(Plate 5, BR.1963/8)*

the main diesel engines while submerged can be summarised as follows:
- Increased speed while submerged, allowing the overall range of deployment to be expanded owing to submarines reaching patrol areas more quickly.
- The ability to charge batteries without surfacing, reducing exposure to enemy detection.
- Conservation of battery capacity, needed for prolonged submerged running.
- Changing the air within the boat, to prevent it becoming stale.
- Providing better humidity within the boat, to help habitability for the crew.

As originally fitted to the A-class boats, the snort mast comprised separate induction and exhaust masts attached to each other and housed within the casing at the after end of the conning tower on the port side. The mast was raised and lowered as an integral unit by hydraulics. The head of the induction mast was fitted with a ball-type float valve that automatically shut off the mast tube from seawater ingress when the boat inadvertently submerged or was temporarily covered by heavy seas. Very evidently, it was essential to keep the boat level and well trimmed in the longitudinal plane when operating at snort depth.

ABOVE HMS *Andrew* as built. Later in her career in 1959 the *Andrew* starred in the film *On the Beach* (based on the Nevil Shute novel of the same name) playing the role of the fictional US Navy nuclear submarine USS *Sawfish* – the Americans were not prepared to cooperate in the production of the film.

BELOW HMS *Andrew* in rebuilt form, returning through the Straits of Johor in 1964 during service in the Indonesia–Malaysia confrontation. For these operations the *Andrew* was refitted with a 4in deck gun to counter blockade-running junks. This gun, which was fired for the last time in December 1974, is now on display at the Royal Navy Submarine Museum.

There were modifications to the snort arrangements during the life of the A-class submarines. Early on, the exhaust mast was relocated to a fixed position on the after end of the radar mast, so that only the induction mast was raised or lowered. From 1949, A-class submarines started to be fitted with a ring-type float valve at the head of the induction mast instead of the ball-type valve, owing to reliability problems, as exemplified by the fact that *Ambush* suffered severe icing of the ball-type valve while in the Arctic in 1948. The final improvement, introduced in 1956, brought a new periscopic-type induction mast housed in the conning tower aft of the radar mast, the after part of the conning tower fin being extended 3ft (0.9m) to accommodate this modification. The head valve was also improved upon and the water trap was replaced with a helical dryer, which, by creating a centrifugal force, removed water more effectively.

In May 1953 one of the A-class submarines, *Andrew*, became the first submarine to cross the Atlantic submerged, using her snort mast.

Refitting, 1955–60

Following the end of the Second World War, the role of the A-class submarines gradually changed as the Cold War took hold, their primary operational function becoming the monitoring and potential interception of Soviet submarines slipping out of their bases in northern Russia.

Between 1955 and 1960 all A-class

submarines (and some T-class boats) were refitted for this new role, with much of the modification designed to improve underwater speed and to reduce noise levels in order to avoid detection. The work included a complete rebuild of the fore and aft hull sections, lengthening and streamlining of the conning tower and upper decks, removal of external torpedo tubes and the deck gun, and the adoption of improved sonar.

All the A-class submarines – with the tragic exception of *Affray* (see below) – had long and eventful careers as the backbone of the Royal Navy Submarine Service, remaining operational until the late 1960s and early 1970s. Indeed, study of the individual operational history of the *Alliance* later in this chapter will quickly convey the global extent of the activities undertaken by many of these boats. The last A-class submarine to remain in service was *Andrew*, which was decommissioned in December 1974 and sold for scrap in May 1977.

The loss of HMS *Affray*

At 16.00 hours on 16 April 1951, *Affray* left Portsmouth under the command of Lieutenant J. Blackburn, DSC, RN, to undertake a week-long simulated war mission, Exercise 'Spring Train', carrying a reduced crew of 50 plus 25 additional men, mainly officers undergoing training, but also included two marines from the Special Boat Service. Her captain's brief was to drop the marines somewhere off the south-west coast of England (reportedly an isolated Cornish beach) and return during darkness to collect them after their land-based mission.

Affray disappeared during her first night at sea. She made normal contact at 21.00 hours that evening and reported that she was preparing to dive, but was declared missing after failing to make her 08.00 report the next morning. The destroyer HMS *Agincourt* led a fleet of search vessels that eventually totalled 24 ships from four nations, but after three days – the maximum time it was believed the crew could survive on the seabed – the search was scaled back. At one point two of the searching ships picked up a Morse code signal made by tapping on the submarine's hull stating 'We are trapped on the bottom', but this did not help in locating *Affray*.

Two months went by before the frigate HMS *Loch Insh* made sonar contact on 14 June with a sunken vessel near Hurd's Deep, a deep underwater valley off the Channel Islands, an area that had been searched soon after *Affray*

THE FATE OF THE 16 A-CLASS BOATS

No	Name	Date	Fate
1	Acheron	August 1972	Sold to J. Cashmore Ltd and broken up at Newport, South Wales
2	Alcide	1974	Sold to Draper & Sons Ltd and broken up at Hull
3	Alderney	August 1972	Sold to Queenborough Ltd and broken up at Cairnryan, Scotland
4	Alliance	1981	Preserved as memorial and attraction, Royal Navy Submarine Museum, Gosport
5	Ambush	5 July 1971	Sold to T.W. Ward Ltd and broken up at Inverkeithing, Fife
6	Auriga	14 November 1974	Sold to J. Cashmore Ltd. and broken up at Newport, South Wales
7	Affray	16 May 1951	Lost with all hands in Hurd's Deep, English Channel
8	Anchorite	24 August 1970	Sold to West of Scotland Shipbreaking Ltd and broken up at Troon
9	Andrew	4 May 1977	Sold to Davies & Cann Ltd and broken up at Plymouth
10	Aurochs	7 February 1967	Sold to West of Scotland Shipbreaking Ltd and broken up at Troon
11	Aeneas	13 December 1974	Sold to Clayton & Davie Ltd and broken up at Dunston-on-Tyne
12	Amphion	6 July 1971	Sold to T.W. Ward Ltd and broken up at Inverkeithing, Fife
13	Alaric	5 July 1971	Sold to T.W. Ward Ltd and broken up at Inverkeithing, Fife
14	Artemis	1 July 1971	Sunk while refuelling alongside jetty at HMS *Dolphin*, Gosport, Hants
15	Astute	1 October 1970	Sold to Clayton & Davie Ltd and broken up at Dunston-on-Tyne
16	Artful	23 June 1972	Sold to Queenborough Ltd and broken up Cairnryan, Scotland

Note The fates of the two uncompleted boats were as follows: *Ace* – hull used for experimental crush tests, then sold in June 1950 to Smith & Houston Ltd and broken up at Port Glasgow; *Achates* – hull used for experimental crush tests, then in 1947 expended after use as a target.

ABOVE HMS *Affray* was lost with all hands in April 1951.

BELOW HMS *Affray*'s CO, Lt J. Blackburn, DSC, is seen here on the far right when serving in HMS *Safari* in 1943.

disappeared when an oil slick had been sighted there. An underwater camera was sent down and picked up the letters 'y-a-r-f-f-a' – Affray backwards – to immediately confirm the identity of the wreck, which was lying at a depth of 280ft (86m).

Despite investigation by divers, the cause of loss was never established beyond doubt. The divers could find no evidence of collision or damage to the hull or conning tower, all hatches and torpedo doors remained shut, and the two emergency buoys had not been released. The search periscope and radar mast were extended, indicating that *Affray* had been running at periscope depth when she foundered. The only external indication of action taken by the crew was the position of the bow hydroplanes at 'hard arise' and the bridge telegraphs in the 'stop' position.

Divers used a primitive radioactive scanning device to attempt to assess the condition of the interior, but the only information of any significance provided by this exercise was an indication that the internal valve for the snort mast was in the 'open' position, suggesting that at least one compartment may have flooded and caused the submarine to sink. Some experts have speculated that the snort mast may have dropped below the surface and suffered some form of jamming or failure of its float valve, allowing water to rush into the submarine; the snort mast was the only part of the submarine to be recovered and it was later found to have been of faulty manufacture. But none of this amounts to conclusive evidence and the true cause of *Affray*'s sinking, with the loss of all hands, remains unknown.

Alliance in service

The following timeline presents the entire service history of HMS *Alliance*, commission by commission, with dates and events presented in as much detail as records allow.

Pre-commissioning
7 July 1943 *Alliance* ordered.
13 March 1945 Keel laid down by Vickers-Armstrong Ltd at Barrow-in-Furness.
28 July 1945 Launched by Mrs Redshaw, wife of Sir Leonard Redshaw, Managing Director of Vickers-Armstrong Ltd, after a period of

two years and two months' construction, completion and initial fitting out.

First commission

11 March 1947 Commissioned into service under the command of Lieutenant K.H. Martin DSC and sailed to Rothesay to commence standard RN acceptance trials.

14 May 1947 *Alliance* officially handed over to Lieutenant K.H. Martin and sailed for 'crew shakedown' and 'work-up' exercises in the English Channel. Her first deployment was to undertake 'snort' trials to record information about conditions related to extended snort patrols.

1 October 1947 Sailed from Portsmouth to Gibraltar, where she took on fuel and water before making passage to a planned position south-east of the Canary Islands.

9 October 1947 Dived to begin the long snort test and remained submerged for 30 days. The assigned on-board naval doctor recorded details concerning the health and morale of the crew. Examinations took place daily if personnel so chose and all were given vitamin E tablets twice daily to counteract 30 days of darkness. One case of appendicitis could have aborted the trial, but fortunately the crewman recovered. Contrary to normal practice, smoking was banned while dived during this trial, and the men resorted to eating Rowntree's fruit gums as an alternative! One crew member, signing himself 'Able Seaman R. Smeaton of Lt Martin's Under-Sea Circus', later wrote to Rowntree's to say how much the product was appreciated under the conditions. During this period

ABOVE *Alliance* on trials off Barrow, August 1946.

BELOW *Alliance* alongside at Freetown, Sierra Leone, after snorting trial patrol in 1947, showing 'bandstand' and after casing with engine room hatch.

ABOVE *Alliance* diving.

LEFT *Alliance* alongside, showing a close-up of bridge starboard side, 4in gun and mast standards. Note the foot holes for mounting the structure from the casing.

Alliance dived deeper than normal to take a bathythermograph recording.

8 November 1947 After covering 3,193 miles fully submerged, when *Alliance* eventually surfaced her hull was covered in marine life. She then proceeded eastwards towards Africa, calling in at Freetown, Sierra Leone, before returning to England.

13 December 1947 Arriving back at Portsmouth, the boat remained operational in home waters.

Second commission

1 March 1948 Recommissioned under Lieutenant Commander J.D. Martin DSC, *Alliance* remained in home waters on exercises and made courtesy visits to Rotterdam, Copenhagen, Odense, Bergen and Stavanger, during the course of this intinerary becoming the first British submarine to make passage through the Kiel Canal after the Second World War.

22 December 1948 Paid off at Chatham for refitting.

1948–49 At Portsmouth for refitting.

LEFT *Alliance* crew members – the 'Bearded Party'. Not shaving saves on-board water supplies!

ABOVE *Alliance* alongside at Aberdeen. The circular aperture visible in the casing provides access to the torpedo loading hatch.

ABOVE RIGHT *Alliance* 'on trot' with another A-boat alongside the depot ship, looking down on to the bridge. Note the differing gun shields.

RIGHT *Alliance* in dry dock during the early 1950s, showing the opening of the bow external torpedo tube, and the bow doors of both internal torpedo tubes below.

Third commission
8 July 1949 Recommissioned under Lieutenant A.J. D'Arcy Burdett, *Alliance* operated with the Home Fleet, visiting Karlskrona, Copenhagen, Aberdeen and Gibraltar.

Fourth commission
20 April 1950 Recommissioned under Lieutenant B.W.M. Clarke.
1950 *Alliance* used during the making of the British feature film *Morning Departure* (released as *Operation Disaster* in the USA) about life in a British sunken submarine. Based on a play by Kenneth Woollard and directed by Roy Ward Baker, the film starred John Mills and Richard Attenborough.
1 November 1950 Paid off and refitted at Portsmouth. Modifications made to snort mast.

Fifth commission
17 July 1951 Recommissioned under Lieutenant Commander R.F. Tibbatts DSC,

ABOVE *Alliance* at sea with crew members on the bridge. Note the ship's crest plaque.

RIGHT A dramatic bows-on view of *Alliance* as built, in floating dock at Portland, showing all six bow torpedo tubes and bow caps. The bow doors of the four lower internal tubes have been removed for maintenance or access.

Alliance operated in home waters, visiting Manchester and Southport.

17 April 1952 As directed, *Alliance* sailed from Fort Blockhouse to the position off Alderney, Channel Islands, above Hurd's Deep, where HMS *Affray* had been lost with all hands on 16 April 1951. A memorial service was held for the crew of *Affray* and afterwards wreaths were cast over the sea.

May 1952–53 No history was recorded for this period.

30 March 1953 Paid off and refitted at Portsmouth.

Sixth commission

15 September 1953 Recommissioned at Portsmouth under Lieutentant Commander C.H. Hammer DSC MBE. *Alliance* continued operating in home waters, periodically with the Home Fleet, with which *Alliance* visited Gibraltar, Marseilles, Nice and Malta before returning to Portsmouth for refitting and maintenance.

May 1954 Deployed with the Home Fleet for 'Operation Loyalty', escorting HM Queen Elizabeth II on the Royal Yacht *Britannia* from Gibraltar homewards after her six-month Commonwealth tour; submarines in company with *Alliance* were *Acheron*, *Anchorite*, *Artful*, *Scorcher*, *Selene*, *Sleuth*, *Subtle*, *Trespasser* and *Upstart*.

Seventh commission

14 October 1954 Recommissioned at Portsmouth under Lieutenant Commander A.T. Chalmers DSC; during the commission command was temporarily given to Lieutenant B.A. Colston.

19 September 1955 Paid off and refitted at Portsmouth for deployment in Canada.

Eighth commission

11 June 1956 After being recommissioned at Portsmouth Dockyard under Lieutenant Commander H.R. Clutterbuck DSC and operating in home waters for a short period, *Alliance* was detached to Canada to join the 6th Submarine Squadron (SM6).

11 September 1956 Arriving at Halifax, *Alliance* joined SM6; after a few days embarking torpedoes and provisions, she carried out a week's trials with HMCS (Her Majesty's Canadian Ship) *Crusader*.

26 September 1956 In company with the 3rd Canadian Escort Squadron, *Alliance* sailed for exercises and a short operational visit to Newport, Rhode Island.

November 1956 Undertook more 'special trials' in the Gulf Stream; these involved snorting in bad weather and working for the listening arrays at CFS (Canadian Forces Station) Shelburne, Nova Scotia. Despite the hard work involved, this was considered monotonous. Trials followed with the RCN (Royal Canadian Navy) and RCAF (Royal Canadian Air Force). *Alliance* then sailed for Bermuda with the 1st and 3rd Escort Squadrons, undertaking basic training en route.

December 1956 Spent a week carrying out anti-submarine training out of Bermuda with the 1st and 3rd Escort Squadrons, followed by a Joint Tactical Course for the Joint Maritime Warfare School. *Alliance* then returned to Halifax to give Christmas leave to those of her ship's company who were RCN personnel – about a third of the crew – and to carry out planned maintenance, including a docking.

January 1957 *Alliance* docked at Halifax.

25 January 1957 Undocked and sailed for Guantanamo Bay Naval Base, Cuba, and participation in Exercise 'Springboard'.

30 January 1957 Though the plan had been to run submerged – and snorting – for much of her passage south to Guantanamo Bay, bad weather forced *Alliance* to complete her journey on the surface. Much delayed, she was only allowed 24 hours in harbour before sailing for five days' trials and passage to St Thomas, one of the US Virgin Islands. On arrival *Alliance* berthed alongside USS *Bushnell*, a submarine depot ship that gave her all possible assistance with stores, victuals and repairs.

March 1957 *Alliance* visited Jamaica.

June 1957 *Alliance* visited St John, another of the US Virgin Islands.

8 August 1957 Annual harbour inspection carried out by 'Cancomflt' (Canadian Commander Flotilla).

9 August 1957 *Alliance* has sea inspection carried out by Commander SM (Submarines).

2 December 1957 Sailed from Halifax for exercises.

11 December 1957 *Alliance* grounded some 50 miles up the coast from Halifax; with exercises aborted she made for base.

13 December 1957 Returned to Halifax, where a Board of Inquiry was convened. *Alliance* was put into dock and surveyed for damage caused by the grounding, with necessary repairs undertaken. The boat had a leak in the radar mast that was successfully repaired and her whip antenna aerial had broken off near its base after nine months' use. This, according to the records, was by far the longest period such an antenna had lasted in this squadron.

22 December 1957 The daughter of crewman Harry Hughes and his wife Minna was baptised Gwyneth Jaqueline in the *Alliance*'s control room by Padre Peglar.

21 January 1958 Not operational since the grounding, *Alliance* was cleared to dive to 170ft and sailed for Exercise 'Springboard' one day behind the fleet.

1/2 February 1958 In company with HM Submarine *Alcide*, *Alliance* spent this first weekend of February alongside the American submarine tender USS *Gilmore* at St Thomas, Virgin Islands, later joined by HM Submarine *Amphion*, which was the first A-class boat built.

3 February 1958 In company with *Alcide* and *Amphion*, *Alliance* sailed for exercises off Puerto Rico with the 1st and 3rd Escort Squadrons, which were en route to Mayport, Florida.

8 February 1958 *Alliance* and *Alcide* detached for a four-day recreational visit to Nassau, Bahamas. Despite a local strike, USS *Gilmore* gave both submarines every assistance. Nassau gave a great welcome, inviting both commanding officers to stay at Government House for the duration of their visit, and both ships' companies were also well looked after.

12 February 1958 Sailed from Nassau for Halifax.

16 February 1958 Arrived at Halifax, completing her last task with the 6th Submarine Squadron (SM6) and the RCN.

22 February 1958 After 18 months on the Canadian station during which she had covered 41,000 miles, *Alliance* sailed for Britain.

4 March 1958 Arrived at Portsmouth a day late after being delayed by bad weather in the North Atlantic. Here she remained until transferred to Devonport (Plymouth).

11 June 1958 Paid off at Devonport for refitting.

October 1958 Brought into Devonport Dockyard for an extensive refit.

ABOVE *Alliance* in dry dock after rebuild. Both external torpedo tubes have been fully removed to accommodate the new sonar array system and its dome. (Compare this image with the pictures of *Alliance* in dry dock in the 1950s on pages 33-4.)

BELOW *Alliance* at sea with the sub-lieutenant OOW (Officer of the Watch) on the bridge taking a compass bearing.

Extended refit 1958–60

Alliance was extensively modernised to meet the demands of the Cold War. As submarines now needed to be quieter and less detectable, the hull was streamlined to reduce noise created by water turbulence. For similar reasons, the conning tower was modified and heightened to 26ft 6in, providing the sleek aluminium 'fin' seen on the boat today.

By now considered unnecessary, the deck gun and external torpedo tubes were removed. Fortunately the original gun access hatch was retained, allowing *Alliance* to be equipped once again with a small-calibre deck gun while serving in the Far East during the Indonesian confrontation in the 1960s.

Another improvement was to the wireless transmitting aerial, which was now supported on a frame at the after end of the fin. This was later replaced with a more sophisticated whip aerial (fitted to the starboard side) that could be rotated hydraulically to a horizontal position.

Ninth commission

21 April 1960 Recommissioned under Lieutenant Commander M.L.T. Wemyss. After undertaking post-refit trials, *Alliance* was transferred to the 3rd Submarine Squadron (SM3) based at Faslane on the Clyde and there she remained, operating in home waters.
13 November 1960 Transferred to the 2nd Submarine Squadron (SM2) at Devonport (Plymouth), operating in home waters.

Tenth commission

1 May 1961 Recommissioned under Lieutenant Commander C.A.W. Russell with the 2nd Submarine Squadron, *Alliance* continued operating in home waters. In this same month the pennant numbers given to British submarines were changed; all existing submarines completed after the Second World War were now numbered from S01 upwards, so *Alliance* had her original pennant number, P417, replaced with S67.
May 1961 Operating out of Devonport with the 2nd Submarine Squadron, during this period *Alliance* visited Le Havre.
22 January 1962 *Alliance* paid off at Chatham for refitting.

Eleventh commission

4 February 1963 Recommissioned under Lieutenant Commander A.G.A. Pogson for service in home waters.

April/May 1963 *Alliance* carried out the first of the class TCSS 6 and Mk XXIII torpedo trials. During this period *Alliance* received orders to sail from Chatham for the Far East station, the global theatre for which she had been specifically designed, as a consequence of the confrontation that had developed between Indonesia and Malaya. In the previous year, 1962, Britain had decided to unify its colonies in Borneo with Malaya to create the country Malaysia, but the Indonesian government, believing that Malaysia would become a puppet of the British after independence, opposed the concept. An undeclared war, with Borneo its focal point, lasted until 1966.

May 1963 *Alliance* first sailed for South Africa, calling at Dakar en route and carrying out sub-surface equatorial current trials on the Equator. *Alliance* next participated in exercises with the South African Navy, which had no submarines.

14 August 1963 After visiting Cape Town, Simon's Town, Port Elizabeth and Durban, *Alliance* sailed for Aden.

5 September to 2 October 1963 Took part in Exercise 'FOMEASWEX III' (Flag Officer Middle East Anti-Submarine Warfare Exercise) in the Arabian Sea with the French Navy.

10 October 1963 With new orders related to the Indonesia-Malaysia conflict, *Alliance* sailed for Singapore to join the 7th Submarine Squadron (SM7).

November 1963 Joined SM7 at Singapore and began an intermediate dry-docking and maintenance period.

15 December 1964 Maintenance completed, *Alliance* commenced 'war patrols' protecting the east coast of Malaya and supporting covert operations along the coastline of Borneo. Missions involved landing Royal Marine SBS (Special Boat Service) and Army SAS (Special Air Service) men together with Royal Marine commandos on the shores of Borneo, landings always carried out when there was no moon. Other HM Submarines attached to the SM7 boats involved were *Ambush*, *Amphion*, *Andrew* and *Oberon*. As a necessity of war, all these

ABOVE *Alliance*, Senior Rates on deck, Far East 1963. Note that the crew are wearing sarongs for coolness; one man sports colourful cotton 'nix', which were commonly bought from Chinese tailors by Royal Naval ratings serving in the Far East.

BELOW *Alliance* engine room crew, November 1964. Few 'donk-shop' 'clankies' wore overalls in the Far East because of the extreme temperatures.

ABOVE *Alliance* with deck gun during service in the Indonesia–Malaysia confrontation in 1965.

BELOW *Alliance* in her new camouflage paint scheme, 1965.

submarines, including *Alliance*, had been fitted with small-calibre deck guns.

1964 After carrying out another intermediate docking at Hong Kong, *Alliance* sailed to take part in various SEATO (South East Asia Treaty Organisation) exercises during the year.

15 December 1964 Recommissioned under Lieutenant Commander John P.A. Purdy. (Note: the author attended the training establishment HMS *Ganges* where Purdy was his NEDO (New Entry Divisional Officer).)

April/May 1965 *Alliance* involved with inshore patrols off Sumatra, acting as a scout where Indonesian marines were known to be active. Besides intelligence gathering and surveillance, agents were being landed on the Indonesian islands, with Royal Marine commandos and SBS and Army SAS actively involved.

September 1965 *Alliance* visited Hong Kong and took part in Exercise 'FOTEX 65' (Photograph Exercise). Submarine escape and re-entry trials from depths of 30ft, 50ft and 80ft were carried out. *Alliance* was also used for trials to test the feasibility of a new submarine camouflage paint scheme.

25 August 1966 Command temporarily under Lieutenant Commander B.I. Nobes RAN (Royal Australian Navy).

28 August 1966 Recalled home, *Alliance* sailed for Britain via the Suez Canal, calling in at Aden, Malta and Lisbon en route during September.

September/October 1966 After visiting Falmouth, *Alliance* arrived at Plymouth and went into Devonport Dockyard for refit.

January 1968 *Alliance* visited the port of Lorient, Brittany, France.

13 January 1968 *Alliance* ran aground on Bembridge Ledge off the Isle of Wight while attempting to come into Gosport. Efforts were made to refloat her on the next high tide but one of the cables snapped.

15 January 1968 *Alliance* refloated by the

lifting vessel *Goldeneye* with help from the tugs *Kinloss* and *Samson*.
February 1968 *Alliance* visited Manchester.
March 1968 *Alliance* visited Rotterdam.
August 1968 *Alliance* visited Gibraltar and Cardiff.
October 1968 *Alliance* visited Den Helder (in the Netherlands) and Gibraltar.
10 November 1968 While submerged at 400ft (121.9m) fire broke out in the motor room, forcing the boat to carry out an emergency surface to tackle the fire and ventilate. The fire was caused by a burst armoured hose, resulting in considerable electrical arcing on the port switchboard. Ratings EM Double and ME Avis prevented more serious damage and received commendations from FOSM (Flag Officer Submarines). (EM = Electrical Mechanic and ME = Mechanical Engineer.)
1–22 January 1969 *Alliance* and *Ocelot* at Newcastle for a courtesy visit.
10 February 1969 *Alliance* paid off at Chatham for refitting.

Twelfth commission
10 November 1969 Recommissioned under Lieutenant Commander C.A.B. Nixon-Eckersall.

Thirteenth commission
9 May 1970 Re-commissioned at Chatham under Lieutenant Commander R.S. Forsyth and began sea trials.

ABOVE LEFT *Alliance* aground on Bembridge Ledge off the Isle of Wight on 13 January 1968.

ABOVE A salvage operation is under way to tow off *Alliance* and refloat her. After 40 hours aground she was towed off by the salvage vessel *Kinloss* (right) after which she returned to Portsmouth under her own power.

June 1970 Sailed to Faslane for further trials, checks and an inspection, then sailed to Devonport to rejoin the 2nd Submarine Squadron to undertake more maintenance.

LEFT *Alliance* at sea viewed from another submarine.

September 1970 Visited Port Talbot and then returned to Plymouth, calling in at Falmouth en route.
October 1970 Leaving Plymouth, *Alliance* then sailed for Portsmouth, then to the Mediterranean, visiting Gibraltar and Malta.
November 1970 While in Malta, *Alliance* took part in Exercise 'Lime Jug' after which she arrived at the NATO Supreme Allied Commander Atlantic Anti-Submarine Warfare Research Centre (SACLANTCEN) at La Spezia, Italy, to take part in anti-submarine warfare (ASW) trials.
December 1970 After visiting Gibraltar, sailed to Devonport (Plymouth) for Christmas leave and maintenance.
January 1971 After passing harbour inspections at Devonport, *Alliance* returned to Gibraltar.
February 1971 Having completed her sea inspection at Gibraltar, *Alliance* went on exercise with the American ship USS *Eagle*.
March 1971 *Alliance* returned to Devonport for dry-docking and maintenance.

April 1971 Operating between Devonport and Faslane, *Alliance* carried out Exercise 'Running Scrap' with the minelayer HMS *Manxman* before returning to Devonport for maintenance and leave.
June 1971 Sailed for Den Helder, the Netherlands, visiting London and Harwich en route.
July 1971 *Alliance* made visits to Dartmouth and Plymouth.

Final command
23 August 1971 Returning to Devonport, command was handed over to Lieutenant A.D.E. Pender-Cudlip. Began trials with HM Submarine *Grampus* during which the two boats visited Falmouth.
August 1971 After returning to Devonport *Alliance* sailed for Norway.
September 1971 After visiting Haakonsvern and Odda, *Alliance* sailed to Rosyth. Leaving Rosyth, she carried out exercises with the aircraft carrier HMS *Ark Royal* and then returned to Portsmouth on completion.

ABOVE *Alliance* alongside at Southampton in 1972.

LEFT *Alliance* at HMS *Dolphin* as Harbour Training Boat. Note also the Sea Cadet sail training brig TS *Royalist* in which the author later sailed.

29 September 1971 While operating off Portland, *Alliance* suffered a violent battery explosion created by a build-up of hydrogen resulting from a fault in the ventilation system. One rating (R.A. Kimber) was killed and fourteen crew members were injured; the accommodation was extensively damaged.
October 1971 Returned to Devonport for necessary repairs, battery cell change and maintenance.
1 February 1972 Beginning a trial dive off Plymouth and diving too steeply, *Alliance* hit the seabed at 122ft below the surface. The impact caused flooding in the engine room, but no casualties were sustained and she was able to recover her correct trim.
March 1972 Leaving Devonport, *Alliance* went on patrol in the Firth of Clyde.
April 1972 Completing her patrol, *Alliance* sailed to Faslane for an inspection, returning on completion to Devonport for maintenance.
May 1972 *Alliance* visited Southampton and Dartmouth, then returned to Devonport.
June 1972 At Devonport *Alliance* undertook a harbour inspection and subsequent maintenance.
July 1972 *Alliance* sailed from Devonport to visit Manchester, taking part in Exercise 'Goldfish' on passage before returning to Devonport for maintenance.
August 1972 After making a courtesy visit to Nantes, France, returned to Devonport to participate in Plymouth Navy Days.

September 1972 Carried out full crew preparations for Exercise 'Strong Express', visiting Cardiff on completion.
October 1972 Docked at Devonport in preparation for Exercise 'Britex 72'.
November 1972 Visited El Ferrol, Spain, for maintenance; crew given shore leave.
December 1972 Still in El Ferrol for maintenance and crew shore leave.
January 1973 Took part in Exercise 'Sunny Seas', then visited Lisbon.
February 1973 Still on exercise, *Alliance* visited Funchal, Madeira, and then Gibraltar.
February 1973 Took part in Exercise 'Sarder', visiting Tangier on completion.
14 March 1973 With her active seagoing career ended, *Alliance* returned to HMS *Dolphin*, Fort Blockhouse, Gosport, and was finally decommissioned.

Nuclear dawn and the demise of diesel-electric submarines

Not only did the cancellation of the construction of 30 A-class boats save considerable money at the close of the Second World War, the rise of a new threat from Russia and its Eastern Bloc satellites caused funding to be redirected towards the development of 'air-independent' submarines – boats that did

ABOVE HMS *Meteorite*, ex-German U-1407.

not require air for their propulsion systems and could therefore stay submerged at considerable depth for long periods. In this field Britain pursued German technology based on the salvaged experimental U-boat U-1407, renamed HMS *Meteorite*, which used high-test peroxide as a fuel source for driving steam turbines.

This work resulted in the construction of two 1,120-ton submarines, *Explorer* and *Excalibur*. Unfortunately high-test peroxide

RIGHT USS *Nautilus* (SSN-571) underway.
(US Navy)

fuel proved highly volatile and unstable in operation and these boats, being prone to unexpected explosion, were colloquially called the 'Exploder class'. Moreover their engine rooms, normally unmanned when under way, often had flames emitting from the top of the combustion chamber.

Meanwhile, the USA – the only nuclear power at that time – pressed forward with the development of nuclear-powered, steam-driven submarines. Besides being well advanced in the field of nuclear science, the USA also had the financial resources to pursue this new technology. Added to this was the far-sighted leadership of the four-star Admiral Hyman George Rickover USN. Known as the 'Father of the Nuclear Navy', Rickover directed the original development of naval nuclear propulsion and controlled its operations for three decades, overseeing in this period the production of 23 nuclear-powered surface vessels (aircraft carriers and cruisers) and 200 nuclear-powered submarines, the first of which was USS *Nautilus*, completed in 1955.

During the early post-war period Britain initially continued to develop diesel-electric submarines of more advanced design and instigated the P-class (Porpoise) and O-class (Oberon) submarines to replace the ageing T-class and A-class boats. There were eight Porpoise-class boats built in the period 1956–59 and they stayed in service until the 1980s. The broadly similar Oberon-class boats came into commission from 1960 and served into the 1990s; there were 27 of these, 13 of which served in the Royal Navy and 14 with various foreign navies.

Meanwhile, the Admiralty observed the qualities of USS *Nautilus* during exercises with the Royal Navy's anti-submarine forces and realised that Britain needed its own nuclear-powered submarines. Under the leadership of the First Sea Lord, Admiral the Earl Mountbatten of Burma, agreement was reached with the US to use its S5W reactor, and in 1958 HMS *Dreadnought* – the seventh Royal Navy ship to bear that name – was commissioned from Vickers-Armstrong. Launched by HM Queen Elizabeth on Trafalgar Day in 1960, Britain's first nuclear submarine was commissioned in April 1963 and served the Royal Navy reliably and effectively until 1980.

LEFT HMS *Dreadnought*, Britain's first nuclear-powered submarine, turns to port to come alongside at Faslane Submarine Base with the casing party closed up and ready to tie up the boat with 'springs'. *(Jonathan Falconer collection)*

Chapter Three

Anatomy of *Alliance*

This chapter starts with an explanation of how a submarine works and then goes on to dissect HMS *Alliance* in considerable technical detail, covering her structural layout, general interior arrangements, external tanks, tank blowing and venting systems, high-pressure air system and main telemotor (hydraulic) system.

OPPOSITE A-class sister boat to *Alliance*, HMS *Aurochs* – this is a close-up view of her bridge and gun platform.

TOP ILLUSTRATION

General arrangement drawing showing the layout of internal tanks in the A-class. *(Roy Scorer based on Plate 1, BR.4549/BR.2507/2)*

1. No 5 main ballast tank
2. AIV tank 835gal
3. TOT tank 1,764gal
4. After trim tank 2,203gal
5. Motor room bilge
6. Drain oil tank 1,132gal
7. After reserve lubricating oil tank 1,730gal
8. Forward reserve lubricating oil tank 1,730gal
9. Air space
10. No 5 fuel tank 4,024gal
11. Snort drain
12. Oily bilge
13. Slop drain tank 168gal
14. No 8 fresh water tank 318gal
15. No 2 battery tank
16. No 7 fresh water tank 500gal
17. No 5 fresh water tank 484gal
18. No 6 fresh water tank 450gal
19. Sewage tank 236gal
20. Telemotor storage tank 219gal
21. 'R' port compensating tank 1,965gal
22. 'R' starboard compensating tank 2,140gal
23. Washing water tank 626gal
24. No 4 fuel oil tank 1,948gal
25. No 4 fresh water tank 435gal
26. No 1 battery tank
27. Distilled water tank 315gal
28. No 3 fresh water tank 578gal
29. No 2 fresh water tank 421gal
30. No 3 oil fuel tank 1,795gal
31. No 1 fuel oil tank starboard 4,465gal
32. No2 fuel oil tank port 4,465gal
33. Forward trim tank 3,026gal
34. TOT tank 2,240gal
35. AIV tank 982gal
36. No 1 main ballast tank

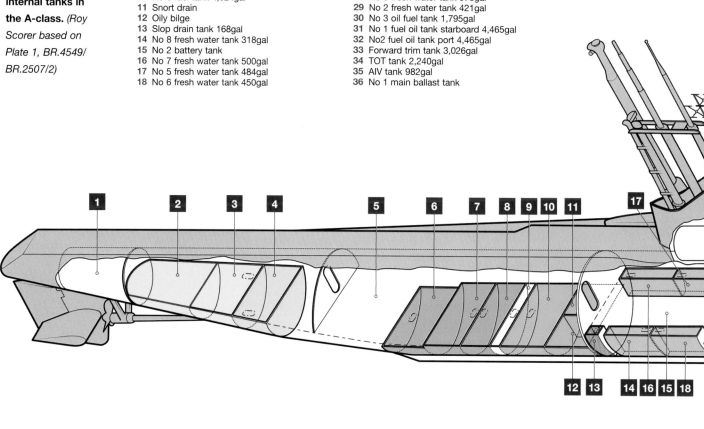

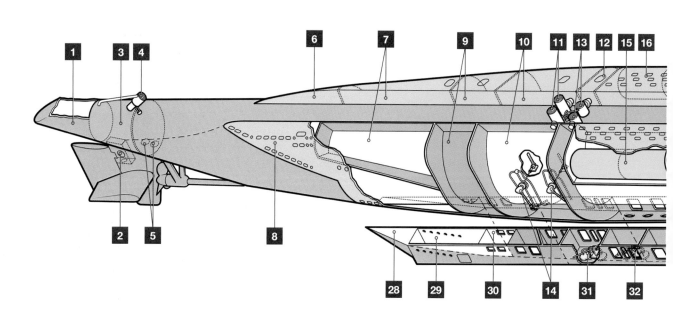

BOTTOM ILLUSTRATION

General arrangement drawing showing the layout of external tanks in the A-class.

(Roy Scorer based on Plate 1, BR.4549/BR.2507/2)

1 Free flood space
2 Free flooding pocket
3 No 5 main ballast tank (MBT)
4 No 5 main vent
5 No 5 MBT free flood holes
6 Free flood space
7 No 4 external oil fuel tank
8 Free flood space
9 No 3 external oil fuel tank
10 No 4 MBT
11 No 4 MBT main vents
12 No 3 MBT
13 No 3 MBT main vents
14 Kingston valve operating gear
15 'O' starboard compensating tank 2,388gal
16 'O' port compensating tank 2,388gal
17 No 2 MBT main vents
18 No 2 MBT
19 No 1 external oil fuel tank
20 No 2 external oil fuel tank
21 Free flood space
22 Gearing for hand-operated bow buoyancy tank vent
23 No 1 MBT main vent
24 No 1 MBT
25 No 1 MBT free flood holes
26 Free flood space
27 Bow buoyancy tank
28 Cast iron block
29 Ballast box
30 Cast iron ballast
31 No 4 port MBT Kingston valves
32 No 4 starboard MBT Kingston valves
33 Radar mast well
34 After periscope well
35 Forward periscope well
36 'Q' tank Kingston valve compartment
37 'Q' tank with float gauge
38 'Q' tank siphon compartment
39 Echo sounding transceiver
40 Cast iron ballast blocks
41 Ballast box
42 Dome casting
43 Asdic dome

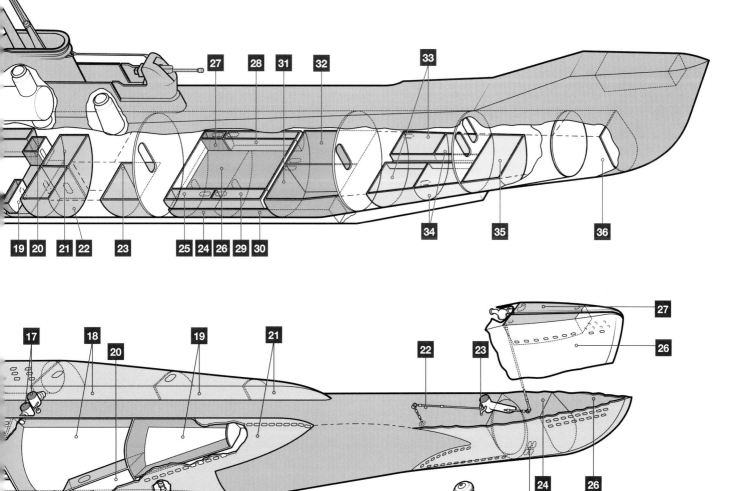

47
ANATOMY OF *ALLIANCE*

How a submarine works

Buoyancy

A submarine is able to operate in the manner it does by control of the natural phenomenon of buoyancy. There are three forms of buoyancy: positive, negative and neutral.

Positive buoyancy Under normal circumstances a ship or submarine has what is termed positive buoyancy when floating on the surface of the sea. In this situation the proportional mass weight of the vessel is in equilibrium to the mass weight of water displaced by the object. This principle was discovered by Archimedes, the Greek mathematician, physicist and engineer who wrote, 'Any body wholly or partially immersed in a fluid experiences an up thrust equal to, but opposite in sense to, the weight of the fluid displaced.'

Negative buoyancy This is the term given to a vessel when the balance between mass weight and water displaced becomes disproportionate, with the result that gravitational forces override all other acting forces and cause the vessel to sink.

Neutral buoyancy This is the condition in which a physical body's density is equal to the density of the fluid in which it is immersed. This offsets the force of gravity that would otherwise cause the object to sink. An object that has neutral buoyancy will neither sink nor rise and thus is in a state of equilibrium.

It is the control of the balance between

BELOW Flood (top) and surface (bottom). *(Roy Scorer)*

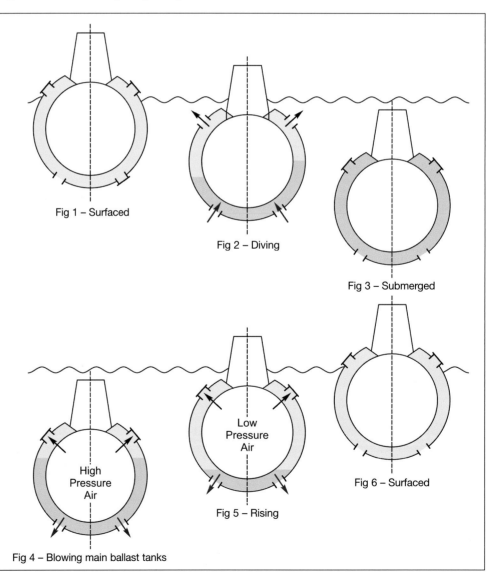

Fig 1 Surfaced – ballast tank main vent valves shut, holding air in main ballast tanks keeping submarine at positive buoyancy.

Fig 2 Diving – ballast tanks main vent valves open allowing air in the main ballast tanks to escape by displacement of seawater entering into tank through free-flood holes at bottom. The displacement weight change creates a negative buoyancy effect, submerging the submarine.

Fig 3 Submerged – main vents shut, main ballast tanks full of seawater. Submarine at neutral to negative buoyancy.

Fig 4 Blowing main ballast tanks – main vents remain shut creating a back-pressure effect when main ballast tanks are blown with high-pressure air, forcing seawater out of free-flood holes. The displacement weight change creates a positive buoyancy effect to surface the submarine

Fig 5 Rising – main vents remain shut. As the submarine rises, main ballast tanks are blown using low pressure blower to eliminate remaining seawater through free-flood holes, creating total positive buoyancy.

Fig 6 Surfaced – main vents shut. Main ballast tanks completely full of air. Submarine is fully buoyant.

positive and neutral buoyancy that provides a submarine with its operational ability to dive or resurface. This control is maintained by changing the weight of the submarine by flooding or emptying its ballast tanks with seawater. In effect the submarine 'hangs' in a state of submergence using a combination of its propulsion unit driving it forward and its control surfaces for manoeuvring, diving and surfacing the boat as desired (ie rudder and hydroplanes).

Stability

The stability of any seagoing vessel is determined by the positions of the centre of gravity, the centre of buoyancy and the metacentre. Centre of gravity is calculated as the point where the total sum of all weights of the submarine act at a single determined point on the centreline, preferably calculated low down in relation to the cross-sectional shape of the hull. The metacentre is the point of intersection between two vertical lines, one line through the centre of buoyancy of the submarine's hull when in a state of equilibrium and the other line through the centre of buoyancy of the hull when the submarine is inclined to one side.

Metacentric height is defined as the distance between a vessel's centre of gravity and its metacentre, and is an initial measurement of static stability; the greater the metacentric height, the slower the rate of roll of the vessel's hull. A simple analogy is to think of a mechanical metronome: the movable weight on a metronome's pendulum rod is slid upwards to decrease tempo and downwards to increase tempo. The greater the metacentric height, the greater the vessel's righting moment – which improves initial stability of the vessel against overturning. Like a metronome, however, a lower metacentric height quickens the natural period of roll of a hull, consequently making a vessel more uncomfortable for those within, especially in the case of a submarine running on the surface.

There is, however, a marked difference between a submarine and a ship once a submarine has submerged. As a consequence of its increased bodily weight (because of taking on seawater) the distance between the centre of gravity and the metacentre shortens to the point where it becomes zero, so that the stability of the submarine is in equilibrium and the metacentric height – or righting moment – is zero.

While on the surface submarines have a greater tendency to pitch rather than roll.

Pressure hull and tanks

The principal part of a submarine's structure, the pressure hull is a very strongly built cylindrical, watertight, steel container that protects the crew and operating machinery from the excessive pressures exerted by seawater pressure as the submarine goes deeper.

Main ballast tanks

Built around the pressure hull, but not integral with it, the main ballast tanks hold the large volume of

ABOVE LEFT Upper starboard side of No 3 main ballast tank beside the fin looking forward, showing the position of 'O' starboard compensating tank fitted within it, which has been temporarily removed during restoration.

ABOVE No 2 main ballast saddle tank and fairing to the main pressure hull at centre right. *(Author)*

RIGHT No 1 main ballast tank free-flood holes. *(Author)*

RIGHT No 5 main ballast tank main vent for emitting air from the tank when diving. *(Jonathan Falconer)*

BELOW General arrangement drawing of the main vent valve. *(BR.4549/2507/2)*

ABOVE Nos 3 and 4 main ballast tank access manholes. *(Jonathan Falconer)*

BELOW No 5 main ballast tank high pressure air tank side isolation valve at the after-hull dome end between Nos 5 and 6 after torpedo tubes. *(Author)*

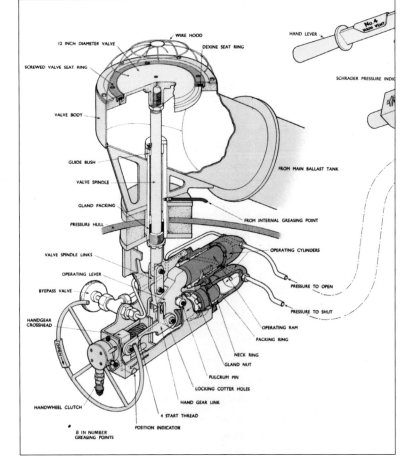

seawater – the ballast medium – that is required to change the bodily weight of the submarine and therefore alter its buoyancy from positive to neutral, permitting the submarine to operate in a state of submergence. Each tank has free-flood holes at its bottom and main vent valves at the top. By the controlled opening of the main vents, all air within the tank is expelled through the vents, displaced by seawater entering up the free-flood holes. Once the tanks are fully flooded, the main vent valves are shut in preparation for 'blowing out ballast' to return the boat to the surface.

In order to surface, ballast tanks are emptied by high-pressure air entering the tanks via controlled blow valves, the expanding air blowing the seawater out through the free-flood holes. The change in displacement by the expulsion of the ballast water – ie, the change in bodily weight – alters the buoyancy state from neutral to positive, bringing the submarine to the surface. Theoretically ballast tanks are not subjected to any change of external seawater pressure at any depth.

Compensating tanks

These are primarily used to maintain the balance of bodily weight by compensating for other weight lost through factors such as fuel usage and expended weapons as well as consumption of fresh water and provisions. These tanks are commonly filled or emptied by the ballast pump, though there are additional systems that may be involved.

Trim tanks

Fitted to control the level of the submarine on an even keel in the longitudinal plane, these tanks, one afore and one abaft, are integrally linked by a common pipe and reversible pump, so that the water maintained within acts as a movable balance to 'trim the boat', especially when operating at periscope depth or 'snorting'.

Construction

Pressure hull

This is a very strongly built, cylindrical, watertight steel container that protects the crew and operating machinery from the increased seawater pressures exerted as the submarine goes deeper. At the 500ft (c.150m)

LEFT **After dome end of pressure hull showing after hydroplane and rudder actuating rods. Note the blanked off-stud pipes of redundant systems.** *(Jonathan Falconer)*

LEFT **After end of pressure hull seen from overhead the stern fairing.** *(Jonathan Falconer)*

operating depth of *Alliance*, seawater pressure is approximately 250psi (17 bar). Given that pressure hulls are built with a safety margin of factor 1½, the DDD (Deep Diving Depth) is 750ft (c.230m) and consequently the DDDTP (Designed Deep Diving Test Pressure) is 375psi (25.5 bar). Below this depth the pressure hull is expected to fail and consequently become crushed.

The hull is made of 1⅞in (47.6mm) thick high-quality steel plate welded over a series of internal steel L-shaped frames disposed at intervals (stations) 21in (53.3cm) apart. Circular in cross-section throughout, the midship section (between frame numbers 45.5 and 19.5) comprises a cylinder of 16ft (4.87m) internal diameter. At either end of this cylindrical section, the hull tapers to 13ft (3.9m) internal diameter at frame numbers 129.5 and 133.5. From these frame stations, the hull is constructed of ¾in (19.1mm) thick steel plate and tapers to a final diameter of 9ft 6ins (2.89m) at the foremost end and 9ft (2.74m) at the aftermost end. The end bulkheads are of 1¼in (28.6mm) thick steel plate and are dome-shaped for maximum strength. The tapered sections are not symmetrical about a central axis as they are constructed so that the top of the hull forms a straight line from forward to aft.

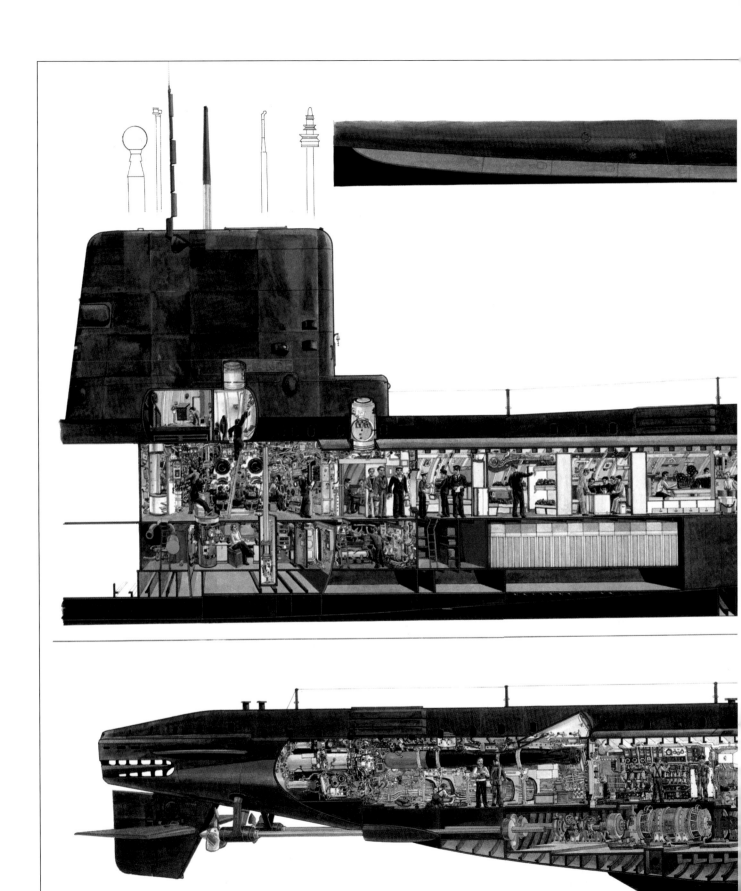

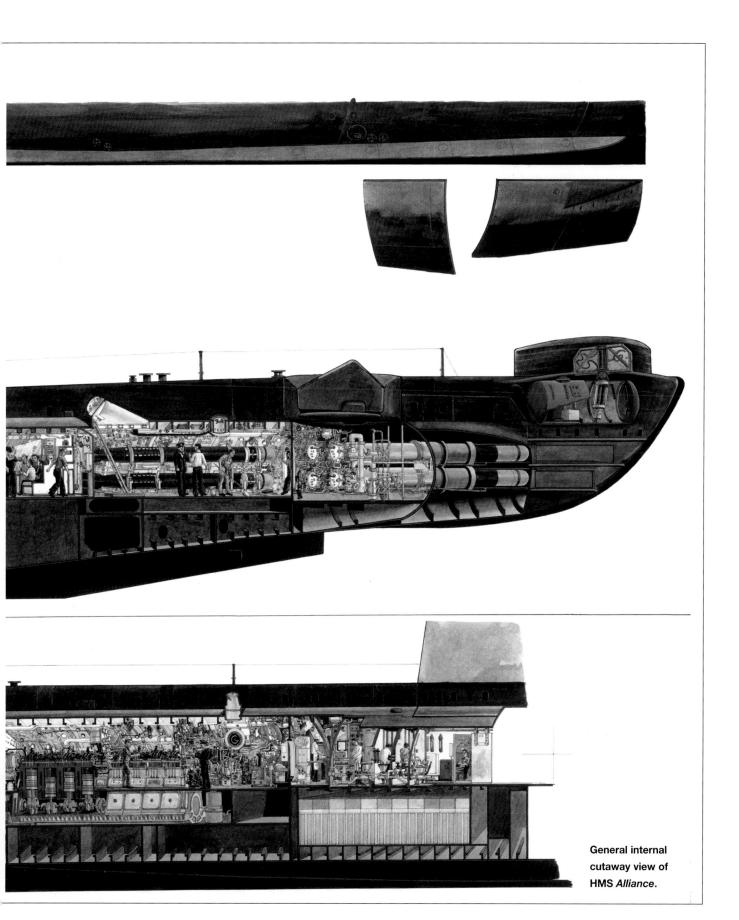

General internal cutaway view of HMS *Alliance*.

RIGHT **Fore part of the box keel (left) and main ballast tank free-flood holes.** *(Author)*

RIGHT **Fore part of the box keel housing the sonar transducers. Note the corroded condition of the ballast tank frames, which have since been refurbished.** *(Jonathan Falconer)*

RIGHT **Pig-iron ingots used for ballast.** *(Jonathan Falconer)*

BELOW **After torpedo loading hatch.** *(Jonathan Falconer)*

Keel

Providing a stable base for docking and grounding, the keel is made of cast iron blocks of rectangular cross-section weighing between five and nine tons apiece. The keel incorporates the ASDIC (sonar) dome and casting, fore and aft ballast boxes, and a fabricated structure incorporating the adjacent main and external fuel tanks; solid-iron ballast may be stowed within this structure. The total weight of all keel blocks and ballast is between 65 and 90 tons.

Hatches and strong-backs

In order to maintain the integrity and strength of the pressure hull, access into it is confined to a limited number of watertight hatches. The only hatches that give access when at sea for operational requirements are the upper and lower conning tower hatches and the upper and lower gun access tower hatches. All are fitted with hinged covers secured by dogged locking rings.

Secondary hatches penetrating the pressure hull comprise two torpedo-loading hatches (one forward, one abaft) and the engine room hatch. All three of these have removable covers that would have been shut and bolted down before the boat proceeded to sea and only opened as required when the submarine was alongside a jetty for loading of torpedoes or for engine room or machinery maintenance. Only

BELOW **Fore torpedo loading hatch (wrongly marked by restoration team). Note pillars for the retractable bollards.** *(Jonathan Falconer)*

under very exceptional circumstances would the torpedo hatches have been opened at sea, if the boat had to surface at sea to load fresh torpedoes from an armament vessel. This risky situation would have been executed extremely quickly and all watertight compartment doors would have been shut and clipped to maintain maximum watertight integrity, the crew remaining closed up within their appropriate watch-keeping compartments for the duration.

The two remaining hatches in the pressure hull are the escape hatches (one in the fore ends, the other in the after ends), each with a hinged, locking cover; for obvious reasons, these hatches are never opened for any purpose other than escape or inspection. Later submarine designs were fitted with segregated escape towers with inner and outer hatches that could be used for exit and re-entry of divers when the boat was submerged.

For obvious practical reasons, the torpedo and engine room hatches are relatively larger than the other hatches and therefore they compromise hull strength to an extent. Consequently, once the hatch cover is bolted shut, this potential weakness is alleviated by the insertion of steel strong-backs that bridge the gap in the circular hull structure at this point. Strong-backs are intentionally made slightly shorter in their bridging length to provide sufficient clearance to absorb stress movement in the hull resulting from the increase in external pressure as the boat descends deeper.

Conning tower and casing

Perhaps the most distinctive feature of a submarine, the conning tower and bridge structure (now often referred to by the American term 'fin') is a working deck or longitudinal platform over the contoured pressure hull, providing the submarine with the facility to 'con' the boat when it is running on the surface. It served the same operational and control function as the bridge of a surface vessel, but with two exceptions:
- Although furnished with a compass, there is no helm position from which to steer the boat; the boat's helmsman was stationed in the control room below, and followed helm orders communicated by voice tube. Most ships of the Second World War, in fact,

had the helm located not on the bridge but instead in a separate wheelhouse further below, helm orders being communicated by voice pipe.
- The conning tower does not contain an engine room telegraph, respective orders being made by voice pipe to the control room; this made it difficult to manoeuvre using individual engine and propeller shafts.

The conning tower also provided a weapons control centre when using the 4in deck gun for a surface action and when undertaking a torpedo attack on the surface.

ABOVE After escape hatch with adjacent reel for the later fitted towed array system. *(Jonathan Falconer)*

General internal arrangements

The internal hull contains six sections divided by five watertight bulkheads.

Fore torpedo room (fore ends)

Located between the fore dome end and watertight bulkhead at frame 26, this space

BELOW Conning tower fin with both periscopes raised and whip aerial. The projection at the fore-foot of the fin encloses the retained gun tower. Also shown is the fin tower access door and fore and aft hand rail used when passing along the casing. The casing drain holes are clearly seen. *(Author)*

ABOVE No 26 watertight bulkhead and door closing off the fore ends torpedo room from the torpedo stowage compartment. Note the main line suction and fire main valves shown at bottom right. *(Jonathan Falconer)*

ABOVE RIGHT No 1 fore torpedo tube (upper starboard) 'shut and clipped' in 2014. *(Jonathan Falconer)*

is 14ft (4.3m) in length and houses the fore torpedo tubes; access is gained through port and starboard single watertight doors that are sufficiently wide for loading torpedoes.

RIGHT Torpedo loading hatch. Note the square sockets for fitting strongbacks. *(Jonathan Falconer)*

FAR RIGHT *Alliance* fore ends showing crew bunks.

Fore torpedo stowage compartment

Extending 24ft 3in (7.4m) between watertight bulkheads at frames 26 and 40, this provides full stowage for reload torpedoes for the foremost torpedo tubes. The lower part is formed of tanks: the fore trim tank and the torpedo-operating tanks. Associated hatches comprise the torpedo-loading hatch and fore escape hatch. Forming the fore escape compartment, the space is furnished with all escape lockers and equipment, built-in breathing systems and fore submerged signal ejector. The space also provides sleeping facilities (hammocks) for additional crew members.

Accommodation space

Extending 45ft 6in (13.9m) between watertight bulkheads at frames 40 and 66, this comprises a longitudinal passage on the starboard side lined with bunks and number 1 battery fuse panel. The port side is subdivided into four messes, the foremost housing an O_2 generator.

Underneath are number 1 battery tank, storerooms and auxiliary machinery.

Control room

This extends 54ft (16.5m) between watertight bulkheads at frames 66 and 97. Often atmospherically recreated in classic black and white movies featuring British actor Sir John Mills, and the more recent epic German U-boat film *Das Boot,* this is the nerve centre of the submarine from which all is operated or controlled.

The foremost section port side contains the officers' wardroom with its pantry in the passageway opposite. The fore part of the control room proper is given over to steering control, gyro compass and engine telegraphs. The entire port side encompasses control stations for both sets of hydroplanes, main ballast tank high-pressure and low-pressure blowing panels, the main vent operating panel and trim pump controls. Further aft on the port side are the wireless telegraphy office, the trim pump and the galley with its associated ovens, hot plates, preparation surfaces, sink and cupboards.

Centrally placed are the eyes of the boat: the attack periscope forward, the search periscope abaft. There are two hatches, one in the wardroom giving entry to the gun access

ABOVE Starboard side passageway and bunks in accommodation space. Note the provision for bunk lights. The mattresses were temporarily removed during the restoration of the boat. *(Author)*

LEFT Control room port side hydroplane control stations (after planes left, fore planes to right) and depth gauges. Note clinometers for 'watching the bubble' indicating 'angle on the boat'. *(Jonathan Falconer)*

RIGHT Wardroom with black heater on after bulkhead, lockers and overhead vent trunking with punkah louvre. *(Author)*

FAR RIGHT Steering control station with rudder angle indicator and gyro repeater providing course steered (top left). *(Jonathan Falconer)*

RIGHT Control Room: main ballast tank blowing panel valves and main vent operating levers. Trim pump control hand wheel and gauge are seen at bottom left. *(Jonathan Falconer)*

BELOW Port and starboard engine order telegraphs. Note the push button telegraph repeater (grey) and red-painted klaxon button, also the clinometer indicating roll angle. *(Jonathan Falconer)*

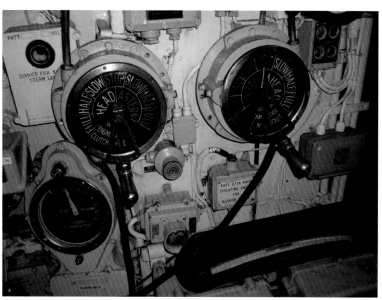

FAR LEFT Entrance to the WT office. *(Author)*

LEFT Fore side of galley. Cooking range hot plates and ovens are underneath. Note the stay bars to retain cooking pots and trays. *(Author)*

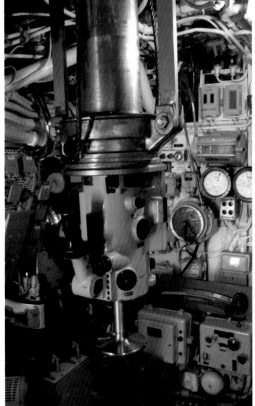

LEFT Control room with attack periscope and helm control station beyond. Note athwartships clinometer indicating 'roll angle on the boat'. *(Jonathan Falconer)*

BELOW Control room looking forward showing search periscope with chart table and sonar to right. At extreme right is pipework for the snort mast drains. *(Jonathan Falconer)*

RIGHT Officers' WC and associated pipework. Note the flexible flushing hose and surrounding extensive electrical cable runs.
(Author)

FAR RIGHT Officers' single washbasin.
(Author)

RIGHT Bathroom for 18 senior ratings.
(Author)

FAR RIGHT Bathroom for 45 junior ratings.
(Author)

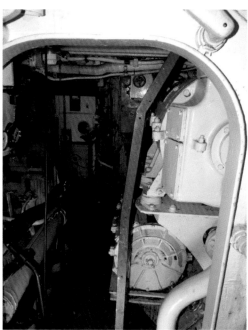

RIGHT Auxiliary machinery space looking forward with entry ladder at centre.
(Author)

tower and the other giving admittance up to the conning tower. The latter leads to the enclosed commander's cabin, which is integrally built within the conning tower, a feature unique to the A-class boats. The starboard side comprises the chart table radar mast well, radar plot, sonar console, the radar office and the snort induction mast system drains. Further aft on the starboard side are four toilets and three segregated washrooms (for officers, senior ratings and junior ratings).

Under the control room is the 'AMS' (auxiliary machinery space) containing the telemotor plant, freon air-conditioning plant, refrigeration plant, distilling plant and a high-pressure air compressor. The remaining space below the control room deck contains the main and ready-use cool rooms, the arms magazines, number 2 battery tank, fresh water and distillate tanks, and slop drain and sewage tanks.

LEFT Engine room looking aft with port diesel (right) and starboard diesel (left), showing respective controls and tachometers (revolution indicators) at the foremost end of each. Note the camshaft covers running along the inner side at deck plate level and the open rocker arm rods that control the inlet and exhaust valve rocker arms. *(Jonathan Falconer)*

Engine room and motor room

This combined compartment is 49ft (14.9m) long and extends between watertight bulkheads at frames 97 and 126. The port side forward corner contains a high-pressure air compressor and diesel fuel oil centrifugal separator. The starboard forward corner contains the ballast (main line) pump and associated six-valve chest, and the main lubrication oil centrifugal separator. Next are the port and starboard diesel main engines with their associated machinery superchargers and main engine clutch. (Of note: the engine room is colloquially named the 'donk shop' in loose referral to the actual engines – donkeys doing the work.)

Abaft the diesels are the port and starboard main electric motors and their associated switchgear and controls. The main motors equally serve as generators for charging the batteries. Following this are the respective tail shaft clutches, propeller shafting and thrust blocks. Under the engines and main motor are circulating water pumps, main lubricating oil pump, lubricating oil tanks, the drain oil tank, fuel tanks and a CO_2 absorption unit.

After torpedo stowage compartment (after ends)

Extending 35ft (10.7m) from the watertight bulkhead at frame 126 and the aftermost dome end of the pressure hull, this compartment

ABOVE Engine room looking aft from the switchboard with engine superchargers in the foreground. Note the shaft order telegraphs for the main motor drive. *(Jonathan Falconer)*

LEFT View looking forward into the engine room and switchboard from No 126 watertight bulkhead at after ends. *(Jonathan Falconer)*

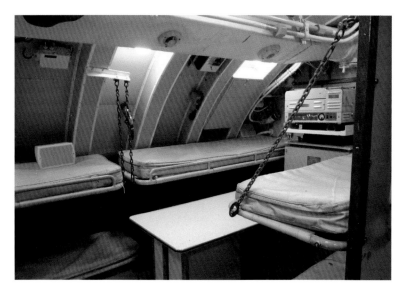

ABOVE **Seamen junior ratings' mess and table, port side.** *(Author)*

contains the two after torpedo tubes and stowage for the after reload torpedoes. The lower part is formed of tanks – the after trim tank and the torpedo operating tanks. Associated hatches comprise the after torpedo loading hatch and fore escape hatch. Also forming the after escape compartment, this space is furnished with all escape lockers and equipment, built-in breathing systems and after submerged signal ejector. Besides containing one bunk dedicated as a 'sick berth', provision is made for hammocks for any additional crew members who may have joined the submarine from time to time. This compartment also contained an O_2 generator.

Accommodation and messes

Living space in all submarines will always be confined. When *Alliance* was initially in service the majority of the junior ratings lived and messed in the fore and after torpedo compartments, slinging their hammocks where possible or 'hot-bunking' in the limited available bunks. Later, the accommodation was modified and it is now divided port and starboard by a longitudinal passage running along the starboard side. This is lined with four sets of bunks in three tiers, two hinged bunks and a urinal at the after end. The port side is subdivided into four messes furnished with seat lockers with two bunks overhead, set in the longitudinal and transverse planes together with a central drop-leaf table. The foremost two messes were dedicated to junior ratings, the first for the seamen, and the second for the 'stokers' (marine engineering mechanics). The next two messes were allocated for senior ratings – the petty officers and the ERAs (engine room artificers). The wardroom, mentioned earlier, was contained in the fore part of the control room section. Both fore and after torpedo compartments could provide sleeping facilities (hammocks) for additional crew members and 'sea riders'.

External tanks

Alliance has ten external tanks comprising five main ballast tanks (numbers 1, 2, 3, 4 and 5), four external fuel tanks (numbers 1, 2, 3 and 4) and a bow buoyancy tank. Constructed of ½in (12.7mm) thick steel plate, these tanks are welded to each side of the keel and extend upwards around either side of the pressure hull. All tanks have watertight access manholes fitted in their upper surfaces.

Main ballast tanks

The function of the main ballast tanks is to achieve, by means of being filled with seawater, neutral to negative buoyancy, enabling the submarine to submerge because the water increases the submarine's bodily weight and consequently its displacement. The reverse function, replacing the water with air, enables the submarine to achieve positive buoyancy in order to rise and resurface. The construction and configuration of the five main ballast tanks is explained below.

Number 1 main ballast tank

Located between frames 10 and 19 this is formed by the foremost circular domed end of the pressure hull and the shape of the submarine's bow, tapering towards the fore bulkhead. Passing through this tank are the four foremost torpedo tubes, and at its aftermost end is the anchor cable locker. For filling, the tank has four (two on each side) oval-shaped, free-flood holes cut into the bottom plating. To expel air, the tank is fitted with two telemotor-operated main vent valves (one on each side). Each tank is fitted with tankside blowing connections from the high-pressure and low-pressure air services, air being the medium used to empty the tanks of water through the free-flood holes. When the main vents are opened the ballast tank will, by virtue of the submarine's dead weight, automatically fill up through the free-flood holes.

Number 2 main ballast tank
Located between frames 56 and 74 this extends around the pressure hull and the box keel at its after end. There are four free-flood holes: one in either section of the keel and two in the underside of both port and starboard tanks close to the keel. Two main vent valves are fitted, one either side and blowing tankside valves. This tank contains number 2 external oil fuel tank.

Number 3 main ballast tank
Located between frames 74 and 90, this is similar to number 2 except for the fact that the

ABOVE No 3 main ballast tank. *(Author)*

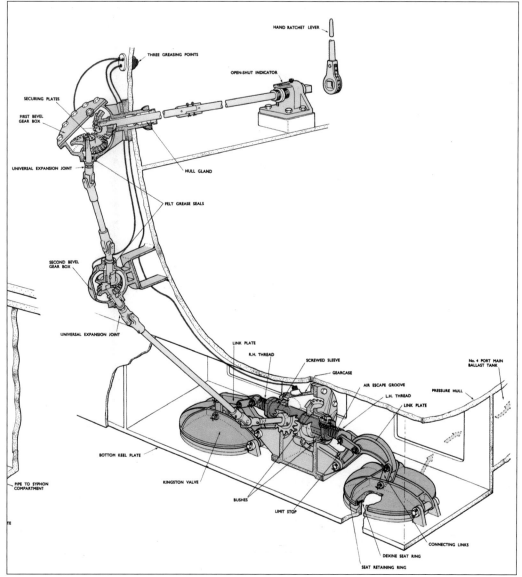

LEFT Hand-operated Kingston valves. *(Plate 2/2, BR.4549/2507/2)*

RIGHT No 5 main ballast tank. *(Author)*

top of the tank is set lower for most of its length in order to accommodate 'O' compensating tanks, which are fitted with telemotor-operated vents and high-pressure and low-pressure air blowing connections. Because number 3 also contains a well to receive the lower ends of the periscopes and masts it has a total of six free-flood holes. This tank is also fitted with an inboard vent leading to the control room.

Number 4 main ballast tank

Located between frames 90 and 100, this is similar to the other main ballast tanks except that the foremost section of the keel is common to the starboard tank and the after section of the keel is common to the port tank. Because these tanks can equally be used for holding fuel oil, there are no free-flood holes in the bottom, so seawater is admitted through hand-operated Kingston valves located in the box keel. Unlike the other main ballast tanks, number 4 is not fitted with high-pressure and low-pressure air blowing services, the absence of which prevents the tank being damaged or over-pressurised when the Kingston valves are shut. The Kingston valves are arranged in pairs in order that both valves are controlled by one single set of rod gearing. The actual valves are set hinged at their fore edge, enabling them to open upwards through 25 degrees, with sea pressure assisting in their opening.

Number 5 main ballast tank

Located between frames 146 and 154, this is symmetrically formed around the centreline, the foremost bulkhead being formed by the after dome end of the pressure hull. Passing through this tank are the two stern torpedo tubes and the operating gear for the after hydroplanes and rudder. Because the telemotor-operated main vent valve is located at the foremost part of this tank, a small-bore pipe is led from the vent to the after end of the tank to prevent air being trapped when the boat inclines 'bow down' to dive. The stern fin and skeg, which houses the rudder, is free flooding. Besides the usual watertight manhole fitted in its upper surface, number 5 has an additional manhole in the stern fin to permit access to the after hydroplane operating gear.

External fuel tanks

All external fuel tanks were pressure-tested to 20psi (1.6 bar) with arrangements provided to equalise internal pressure and sea pressure when the boat dived. The construction and configuration of the four external fuel tanks is explained below.

Number 1 external fuel tank

This is located forward on each side of the hull between frames 43 and 56. There are two vents that are led inboard from the highest points of the tank.

Number 2 external fuel tank

Located between frames 56 and 69, the foremost lower parts of number 2 main ballast tank are divided off to form number 2 external fuel tank, which extends down to the keel, the fore section being common to the port tank, the after part

BELOW Stern fin and skeg. *(Author)*

common to the starboard tank. The tank tops incline upwards towards the bow and are fitted with an inboard vent pipe at the highest point.

Number 3 external fuel tank

Located between frames 100 and 107 this fuel tank, like number 2 main ballast tank, extends the full depth of the external structure and where the keel is divided (station 104), the forward section forming part of the port tank, the after section forming part of the starboard tank. One inboard vent valve is fitted for each tank.

Number 4 external fuel tank

Located between frames 107 and 12, this fuel tank is similar to number 1 external fuel tank, but has long, shallow tanks set high up on the hull, with two vents each.

Bow buoyancy tank

Not an original design feature, the bow buoyancy tank was fitted after initial sea trials in heavy weather had proved its necessity to improve forward buoyancy when surface running. Fitted above the fore torpedo tubes, the tank is 4ft (1.22m) deep with an open bottom constructed to follow the flared shape of the bow. The tank is furnished with vent valves (one to port, one to starboard) operated by rod gearing from inside the torpedo stowage compartment. The tank is self-draining, with no blowing facilities fitted.

Other ballast-related tanks

There are three other miscellaneous external tanks as follows:

'Q' tank

Fitted in the fore part of the keel, this single tank provided a sudden excess of ballast water weight – by means of 'quick flooding' – to drive the fore end of the submarine into a steep dive angle if the boat needed to be submerged rapidly to avoid danger, whether from an attack or a potential collision with another vessel. This action, colloquially called a 'crash dive', was not used as a substitute for main ballast tanks to dive the boat.

'O' compensating tanks

Two separate cylindrical compensating tanks with a capacity of 2,388 gallons each (10,853.5 litres) are fitted inside number 3 main ballast tank, one to port, one to starboard. Used under controlled conditions when dived, these tanks had two functions.

Firstly, they provided additional ballast water to compensate for dramatic or sustained changes to the boat's overall bodily weight, such as when expending torpedoes or after sustained use of fuel oil or consumption of fresh water.

Secondly, fitted port and starboard, they could be used to adjust the transverse trim of the boat. Connected to the main line system, both are filled or discharged using the ballast pump or blown by air.

BELOW 'Q' tank and related services. *(Plate 4/2, BR.4549/2507/2)*

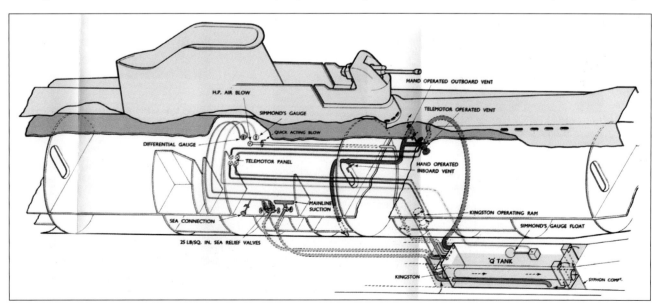

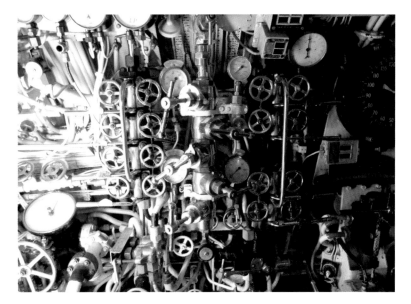

ABOVE **Main blowing panel with master blow levers centre and main ballast tank main vent operating levers to left and right.** *(Author)*

BELOW **No 2 main vent, starboard. Note steel cotter pin adjacent. This is inserted to prevent inadvertent opening.** *(Author)*

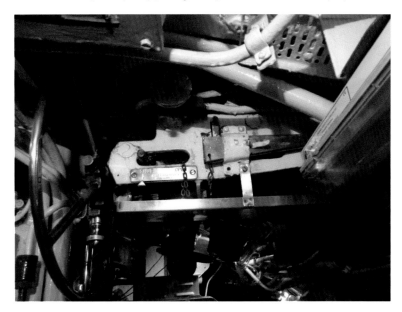

Operation of blowing water from 'O' port to 'O' starboard.

1. Open 'O' port and starboard main line tank-sides (valves).
2. Open 'O' cross connection.
3. Open 'O' port tank-side blow.
4. Shut 'O' starboard tank-side blow.
5. Build up pressure in 'O' port then shut tank-side blow.
6. Open 'O' starboard tank-side blow and inboard vent.

'R' compensating tanks

Fitted amidships within the pressure hull, these are divided port (1,965 gallons) and starboard (2,140 gallons) and served a similar purpose to 'O comps', being likewise filled and discharged from the main line system. These tanks were also used to provide circulating cooling water for the high-pressure air compressors when operating at lower depths.

Tank blowing and venting systems

Main vents

Under controlled operation, these valves release (vent) air from main ballast tanks when opened, permitting tanks to fill with seawater by means of displacement. Shut when blowing main ballast tanks, these vents enable air pressure to build up within the tank and expel seawater in order to surface the boat.

Eight in number, all main vents are fitted along the top of the pressure hull joined to their respective main ballast tanks by 12in (304.8mm) diameter pipes. Physically each is located outside the pressure hull with its operating gear internally within the boat. All shut upwards in order that pressure in the tanks keeps the valves firmly seated. That fitted to number 5 main ballast tank differs in that it is not mounted vertically and thus requires rods and levers for operation.

All main vents are operated by telemotor (hydraulic) pressure controlled from the telemotor panel in the control room. In the event of telemotor system failure each can be operated locally by engaging the clutched hand gear after opening the hydraulic oil bypass valve. To prevent inadvertent opening (especially when in harbour), cotters, or steel pins, can be inserted to mechanically 'foul' the operating gear.

Main tank blowing arrangements

High-pressure direct blowing system

With the exception of number 4 main ballast tank, all main ballast tanks have connections to the high-pressure air blowing panel located in the control room. The blowing panel is supplied with high-pressure air from any of the four high-pressure air bottle groups at a pressure of 4,000psi (275.8 bar) via a distribution valve chest.

Connections to numbers 1 and 5 main ballast tanks lead from their respective valves on the blowing panel to the combined high-pressure and low-pressure tank-side valves located on the domed ends of the pressure hull. Numbers 1 and 5 main ballast tank blows are each divided into two pipes after leaving the blowing panel, one to each tank-side valve, port and starboard.

Regarding number 3 main ballast tank, two additional blows are fitted, each dividing into two pipes to control the amount of air into each side of the tank. These split blowing valves are integrally housed in the same valve casting. Each are quick-acting, screw-down valves with toothed operating wheels that engage with a centrally mounted gearwheel attached to a control lever. When the lever is in the central position, both tanks are blown equally. When the lever is moved to one side or the other, the blowing valve on that side will fully open and the other will close. Normally operated in the central position, the lever is moved over accordingly to alter athwartship trim of the boat if necessary.

Low-pressure air line and high-pressure direct blowing system

Commencing from the fore end of the boat, this system supplies the following valves or services:
- Number 1 main ballast tank high-pressure and low-pressure tank sides.
- Numbers 1 and 2 low-pressure master blows.
- 'Q' tank high-pressure blow.
- Number 2 main ballast tank, port high-pressure and low-pressure tank sides.
- Number 2 main ballast tank, starboard high-pressure and low-pressure tank sides.
- Number 3 main ballast tank, port high-pressure and low-pressure tank sides.
- Number 3 main ballast tank, starboard high-pressure and low-pressure tank sides.
- Direct high-pressure blowing panel.
- Split high-pressure blow to number 3 main ballast tank.
- Number 4 main ballast tank, port low-pressure tank side.
- Number 4 main ballast tank, starboard low-pressure tank side.
- Numbers 3, 4 and 5 low-pressure master blows.

The system also incorporates a high-pressure to low-pressure blowing valve and can be cross-connected to the low-pressure blower via a blower shut-off valve.

Reduced-air blowing and tank venting systems

Air pressure reduced from 4,000psi (275.8 bar) to 50psi (3.5 bar) is subdivided into three parts: a blowing line, a venting line and a combined vent/blow line, system pressures being further reduced to a pressure of 10psi (0.69 bar) where required. Systems and facilities provided are as follows:
- Numbers 1, 2, 3 and 4 trim tanks – blow.
- Forward trim tank – vent/blow.
- Number 1 main ballast tank inboard – vent.
- Forward torpedo overflow tank – vent/blow.
- Numbers 1 and 3 main ballast tanks (port and starboard) inboard – vent.
- Port forward external fuel group – vent/blow.
- Starboard forward external fuel group – vent/blow.
- Forward internal fuel group – vent/blow.
- Numbers 2 and 4 fresh water tanks – vent/blow.
- Urinal tank – vent/blow.
- Numbers 1, 2, 3, 4, 5, 6, 7 and 8 fresh water tanks – vent/blow.
- Water washing tank – blow.
- Gun tower and conning tower inboard vents – vent.
- 'O' tank outboard – vent.
- 'O' tank inboard – vent.
- 'Q' tank outboard – vent.
- 'R' tank port – blow.
- 'R' tank starboard – blow.
- Trim line – vent/blow.
- Slop drain and sewage tank inboard – vent/blow.
- Lubricating oil tanks – vent/blow.
- Internal fuel tank – vent/blow.
- After external fuel tanks port and starboard – vent/blow.
- Main engine circulating water system – blow.
- After internal fuel tank – vent/blow.
- After external fuel group starboard – vent/blow.
- After trim tank – vent/blow.
- After torpedo overflow tanks port and starboard – vent/blow.
- Numbers 7 and 8 trim tanks – blow.

Low-pressure blower

All of the main ballast tanks on A-class submarines can be blown with air provided from a Reavell motor-driven low-pressure blower, the sole function of which is to bring the submarine to full buoyancy after surfacing. Located in the auxiliary machinery space (AMS), the blower discharges free air at 1,000cu ft per minute at a pressure of 11–15psi (0.76–1.03 bar) into the main ballast tanks rather than exhausting the high-pressure air system and its bottle groups after the initial blowing of tanks. As a secondary function it can be used for compartment testing and salvage blowing when the boat is on the surface.

By virtue of the fact that the blower draws in its air from within the boat, subsequently creating a vacuum, this facility was not to be run until the conning tower hatch had been opened and was never to be run with the engine room hatch watertight door shut, to prevent it competing with the engines for air when the diesels were running.

A 4½in (11.4cm) bore discharge pipe from the blower leads through the underside of a valve chest located in the wardroom and thence to a second valve chest outside the wireless room. Pipes run from the forward valve chest to numbers 1 and 2 tank sides and from the after valve chest to numbers 3, 4 and 5 tank sides, each fitted with a master blow valve on the respective valve chest. Pipes from the master blow valve to the tank sides are of 2in (5.1cm) diameter. Compartment pressure-testing blanks are fitted to the low-pressure air line adjacent to each bulkhead. A length of copper pipe was carried within the boat for connecting the air line to the inter-bulkhead communication voice pipes at these positions to enable compartments to be blown for salvage reasons.

The blower comprises a rotor eccentrically mounted within a casing with six radial slots containing a vane cut in the rotor, the vanes being free to move in and out as the rotor turns on its axis. The gap formed between the rotor and casing is minimum at the top and maximum at the bottom. As the rotor rotates, each vane moves outwards, under centrifugal force, as it travels from the top to the bottom of the casing, and then moves inwards again when returning to the top. When extended, the vanes divide the space between rotor and casing into segments, the volumes of which increase as the vanes move from top to bottom and decrease as they move back to the top (in effect creating a rotary reciprocal action). Air is drawn through the blower inlet as the volume increases and discharged through the blower outlet as volume decreases. The tips of the vanes are protected from wear by a close-fitting perforated drum inserted concentrically between rotor and casing. When the rotor is revolving the vanes grip the drum, causing it to rotate.

Running the LP blower

The procedure for running the blower is described in this section.

Preliminary checks

1 Check hatch in auxiliary machinery space is fully open.

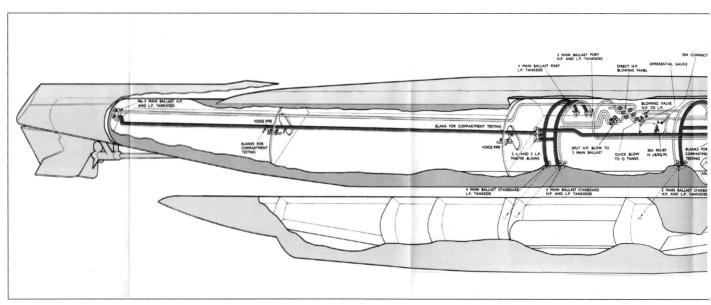

2 Check air intake is clear.
3 Check sight oil level in mechanical lubricator replenishing tank is correct (fill as required).
4 Check drain cock adjacent to discharge valve is open.

Starting the blower
1 Open blower discharge valve.
2 Set starter field regulator to full field (zero on the scale).
3 Press start push button.
4 Make respective knife switches in sequence (1, 2 and 3).
5 When motor is running, the start button can be released.
6 Shut low-pressure air line drain cock.
7 Adjust field regulator to provide motor speed required.

If the blower is run with too few main ballast tanks on the low-pressure air line, pressure will increase to a point where the relief valve will lift. Any increase in pressure will cause the blower to overheat more quickly.

Stopping the blower
1 Open the drain cock.
2 Move field regulator to full field.
3 Brake knife switches in reverse sequence (3, 2 and 1).
4 Shut blower discharge valve.
5 Shut the drain cock.

When the blower rotor has stopped revolving, the drum should be heard to continue rotating for a short while. If this fails to happen, investigate the cause immediately. The probable cause is either ball bearings seizing or vanes sticking. The remedy is to check that the vanes are free; turn the rotor slowly by hand and listen for the blades to fall one after another into the bore of the drum.

ABOVE **Port main motor switchboard knife switch gear.** *(Author)*

High-pressure air system

High-pressure air is supplied at 4,000psi from a total of 15 bottles with a combined volume of 136.5cu ft, individual bottle capacity being 9.1cu ft. Bottles are arranged into four individual bottle groups as follows:
- Torpedo stowage compartment, starboard – three bottles (27.3cu ft).

BELOW **Low-pressure air line and high-pressure blowing system.** *(Plate 4, BR.4549/2507/2)*

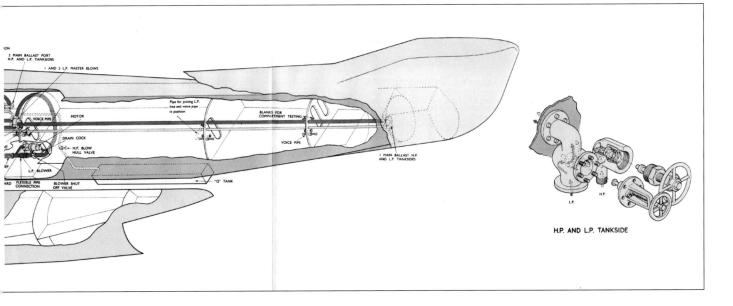

- Torpedo stowage compartment, port – three bottles (27.3cu ft).
- Engineer's store – five bottles (45.5cu ft).
- Auxiliary machinery space – four bottles (36.4cu ft).

Bottle groups all supply the high-pressure air ring main passing through the boat or direct to the main ballast tank blowing panel in the control room. The ring main also supplies various other high-pressure air services as well as 'reducing stations', where air pressure is lowered for other services such as diesel air-start bottles, telemotor systems or air-loaded accumulator air bottles.

The high-pressure air ring main comprises two separate systems, as follows, joined by a cross-connection valve in the control room:

- **Forward ring main** This runs from the fore end of the control room to the after end of the fore torpedo stowage compartment with a leg to the starboard side of the torpedo tube space.
- **After ring main** This runs from the centre of the control room to the forward end of the engine room with a leg passing aft on the starboard side of the after torpedo stowage compartment.

Bulkhead stop valves These are fitted to the ring main on the control room side of each watertight compartment bulkhead. An additional lead is taken from the control room side of the bulkhead stop valve, through a compartment blow valve, to an open-ended pipe in the next compartment.

HP air distribution

The supply from each bottle group leads to the underside of two valves in the main air distribution valve chest, which controls air to the blowing panel and high-pressure air ring main. Valves in this chest are identified A, B, C and D, while those supplying the ring main are numerically identified 1, 2 and 3. The high-pressure air system also supplies the torpedo charging and separator columns.

Bottles in each group are interconnected by copper piping to respective stop valves in the group control chest, comprising five bottle-isolation valves and a drain valve at the lowest point, enabling residue water to be blown out.

All bottle groups are normally recharged by two motor-driven high-pressure air compressors (see below).

High-pressure air compressors

Two Reavell motor-driven high-pressure air compressors are fitted, one in the engine room platform, the other within the auxiliary machinery space. Each compressor discharges air at 4,000psi into the high-pressure air ring main and respective high-pressure air bottle groups. Running at a speed of 800rpm, each compressor has the capacity to charge a reservoir of 15cu ft to 4,000psi in 80 minutes. Although normally run for charging the high-pressure air system when the submarine is surface running, compressors sometimes need to be run while snorting (when they can tend to draw a vacuum within the boat) or while running deep (when it is necessary that they are run using a separate cooling water system).

The driving motor for each compressor is directly coupled to the crankshaft of the mechanical working part of the compressor, the two aligned components jointly sharing a mild-steel bedplate. Air drawn in from the atmosphere is compressed in four stages by means of single-acting pistons; the first- and second-stage pistons work direct from the crankshaft with the third- and fourth-stage pistons integrally attached above them, so that first/fourth and second/third move together. Between each compression stage, air is passed through seawater coolers. Relief valves are fitted on the outlet side of each stage to protect the compressor.

Supplied by an integral water pump for

BELOW Forward HP air compressor. Note the first-stage relief valve and first-stage/second-stage cooler on the left. *(Author)*

cooling, each compressor has a hull inlet valve and hull discharge valve, with additional suction and discharge valves to isolate the compressor. The fore compressor (located in the auxiliary machinery space) has additional circulating water isolation valve connections, fitted to the 'R' port compensating tank, that enable the compressor to be run when operating at deep diving depth without subjecting the air compressor coolers to high sea pressure.

Starting and running a high-pressure air compressor

Prerequisites

1 Prepare high-pressure air bottles to be charged by draining off water and open high-pressure air system to connect bottles to the compressor separator discharge column discharge valve.
2 Check lubricating oil sump is clear of water by opening drain cock.
3 Using dipstick, check oil level in sump is at correct working level; fill as required.
4 Open seawater cooling hull inlet and discharge valve; if deep dived, ensure hull inlet and discharge valves stated above are shut, and open connection valves to 'R' port compensating tank or main engine circulating system.

Procedure

1 Turn compressor over two complete revolutions, ensuring it is mechanically free to turn.
2 Check lubricating oil pressure valve hand wheel is screwed down to ensure good oil pressure is maintained on starting.
3 Open all stage separator column drains.
4 Open separator column inlet valve.
5 Check starter shunt field regulator to full field.
6 Press start button; when compressor is running adjust the shunt field regulator to give required speed.
7 Adjust oil pressure relief valve to read 15psi on the pressure gauge and set the gauge cock so that needle is steady but 'live'.
8 Check circulating water is flowing freely by pressure gauge reading and visually at the vent in the discharge pipe from compressor.
9 Shut all stage drains and air pressure will begin to rise. All the compressed air will now be discharged through drains on the separator column.
10 Shut all drains on the separator column and when pressure rises to be equal to that in the high-pressure bottles, then open the column outlet valve.
12 Adjust separator column fine drain, permitting water to escape without undue loss of air.

While running

1 Open all stage drains frequently to drain off water.
2 Check that mechanical lubricator is delivering oil at two drops per minute.
3 Monitor compressor, ensuring that discharge pressure at each stage is within parameters.

Maintenance

The compressors are generally fully overhauled and water pressure-tested at each designated submarine refit, although the inter-stage coolers should be examined annually. From personal experience, the only major problem that may occur is failure of the neoprene O-ring seals fitted at the end plates of the third- and fourth-stage coolers, often caused by high temperature at these points.

Main telemotor (hydraulic) system

Using oil under pressure as a medium, this hydraulic control system mechanically operates valves, levers, hydroplanes, rudder and other external components that are beyond physical manipulation by the crew.

Air-loaded accumulators

Two motor-driven Imo pumps (see below) deliver oil for the system at a working pressure of 1,250–1,515psi (86.2–104.6 bar). Taking their suction from a replenishing tank, the pumps discharge oil to two air-loaded accumulators where oil is stored under pressure to actuate the telemotor-operated equipment in the system. One air-loaded accumulator is located in the auxiliary machine space, the other at the after end of the engine room. Containing 6,000cu ft of oil pressure tested to 2,250psi (155.3 bar), each air-loaded accumulator comprises a vertical cylinder that houses a free-operating piston; the volume above the piston is connected to two air bottles charged from the high-pressure air system. The volume under the piston is connected to the main telemotor pressure line that extends throughout the boat with isolation valves at each respective component/system it operates, such as the torpedo tube bow cap doors. The 219-gallon (995.4-litre) telemotor oil storage tank is located under the control room abaft 'R' port compensating tank.

The Imo pumps normally run automatically by means of electrical contacts on a pump-control pressure gauge operating 'post office'-type relays that start the pumps when pressure falls below 1,250psi (86.2 bar) and stops them when pressure reaches 1,515psi (104.5 bar). The pump-control pressure gauge is connected to the bottom of each air-loaded accumulator so that the stopping and starting of pumps is directly dependent on oil pressure under the pistons and hence the air pressure above them. When the air-loaded accumulators are discharged, air above the piston expands because working pressure is less than when they are fully charged. Shut-off valves are fitted to the air-loaded accumulators and to oil connections to the gauges so that either air-loaded accumulator, should it become defective, can be fully isolated without detriment to the overall working of the system.

Each air-loaded accumulator has a main cylinder that comprises a hollow, forged-steel body with flanges at top and bottom. The upper flange, made of high-tensile brass, has a hollow indicator cap tube standing vertically from its centre, and coiled around it is a primary and secondary indicator coil enclosed within a watertight brass cover. At the bottom there is a manganese-bronze cover plate fastened to the body flange with 12 steel studs and nuts, the joint comprising a neoprene or dexine ring. The actuating piston, made from gunmetal, has a working clearance within the bore of the cylinder of 0.007in. Needing to be both airtight and oil-tight, it has U-shaped fabric rings fitted back-to-back into grooves around its periphery.

Accumulator air bottles

The system has two high-pressure air cylinders connected to the top of each air-loaded accumulator, complete with air strainers, these cylinders having capacities of 9.1cu ft and 3.75cu ft. Cylinders are charged with air from the high-pressure air ring main, and the charging connection pressure gauge and system relief valve are located in the control room. Each bottle is fitted with its own stop valve with a drain valve fitted in the line for each pair of air bottles.

Telemotor system pumps

Two Mirrlees Roscru Imo pumps are fitted adjacent to their respective air-loaded accumulators (ALA) sited in the auxiliary machine space and engine room. The important advantage of their design is silent, vibration-free operation, due to oil flow being continuous, without pulsation or turbulence. Each pump comprises a positive-displacement rotary screw pump driven by an electric motor sharing a common, resiliently mounted bedplate.

Plant operation

Air pressure in each air-loaded accumulator pushes the piston down its cylinder, forcing the oil under the piston into the telemotor system. When pressure drops to 1,250psi there is approximately 3,000cu ft of oil remaining in the air-loaded accumulator, at which point the pump starts to recharge the system, forcing the piston back up to its point of travel. Should the ALA become completely discharged, owing to a major oil leak, its piston will descend on to the crown-metal seat. With the piston in any other position, oil and air pressure are effectively equal above and below provided that there is no leakage past the piston rings.

To ensure that the system operates correctly by maximising the use of the air-loaded accumulators, it is important that air is maintained at correct pressure and that the air system has no leaks. Switches are fitted in the system to enable the contact gauge to operate pumps individually or together. Pumps can also be manually started or stopped at their starters.

Telemotor system hand pumps

Should a pump fail, the system has two emergency hand pumps, one sited in the forward torpedo stowage compartment, the other in the after ends. Secured to the deck, each hand pump is operated by a manual lever that can be attached to either end of the pump according to requirement: at one end the lever operates a single-acting plunger that delivers a small quantity of oil at high pressure; at the other end the lever operates a double-action plunger that delivers a much larger quantity of oil but at far less pressure.

Auxiliary machinery and equipment

Ballast pump and main line system

Essentially the most important water system in the boat, the multifunctional ballast pump and its associated system has three major uses:
- Transferring seawater as ballast water between 'O' and 'R' port and starboard compensating tanks, and likewise with 'Q' tank, consequently redistributing the bodily weight of the boat.
- Trimming the boat by transferring water between the fore and after trim tanks.
- Pumping out various tanks and discharging oily bilge water and seawater overboard.

The ballast pump is designed for silent operation down to a depth of 680ft. Its discharge capacity depends on depth: at 80ft depth the rate is 33,600 gallons per hour; at 680ft depth the rate decreases to 2,240 gallons per hour.

Designed and manufactured by Messrs Drysdale, the pump is mounted vertically on the starboard side of the engine room. Measuring 89in (226.1cm) high, it comprises, from the top, a rotary air pump, a 29hp DC electric motor with a running speed of 1,880rpm, and a four-stage centrifugal water pump, all sharing a common vertical shaft formed from separate drive shafts rigidly coupled together. The weight of each of the rotating portions is taken by thrust bearings located in the upper bearing housing of the motor. Rotation of the pump is anti-clockwise when viewed from above.

The pump is connected to a 4½in (11.4cm) diameter main line running down the starboard side of the boat. A six-valve suction/discharge chest enables the pump to take suction from the sea and deliver water fore or aft as required, or discharge it overboard. Valve chests connected to the main ballast line throughout the boat enable various internal tanks, including the fore and after trim tanks and the compartment bilges, to be indirectly connected to the pump. The ballast main line is also fitted with hose connections for various other purposes, from firefighting to washing down cables.

Air pump

An air pump is fitted for two reasons:
- To extract air from the suction side of the pump and system, thereby creating a vacuum into which water will pass from the tank into the line and thus prime the pump.
- To maintain vacuum when the pump is running, ensuring maximum efficiency and preventing possible cavitation.

Although the air pump can be switched on and off by opening a cock, it must always rotate when the pump is running.

Water pump

The water pump comprises four impellors mounted one upon another and keyed to a common shaft, each impellor rotating inside its own separate chamber within the vertical cylindrical stator casting.

When the air pump creates a vacuum in the suction line, the water pump is primed by drawing water along the suction line into an air separation chamber and common suction chambers, and thence to the impellors. The centrifugal force set up by the desired speed of rotation causes water to be expelled from each of the peripheral slots in each impellor into the annular discharge passage and thence into the common discharge system. This process creates a vacuum between the impellor blades into which more water is drawn from the suction side of the pump and then discharged, forming a continuous process of discharge and suction into the impellors.

The impellors can be coupled three ways, dividing pump output into differing proportions of quantity and pressure:
- **Four impellors in parallel (A)** All four impellors take their suction in common and discharge passes into a common chamber. By this means a large quantity of water is pumped at relatively low discharge pressure.

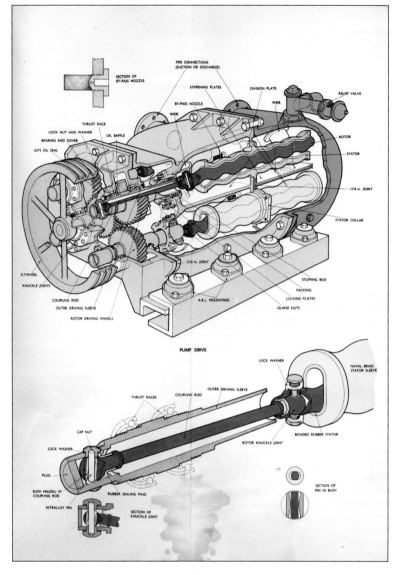

- **Two impellors in parallel (B)** The first and third impellors take their suction from the suction line and discharge respectively to the suction sides of the second and fourth impellors, which in turn discharge into a common discharge chamber. Although the quantity of water pumped is less, discharge pressure is greater than for mode A.
- **Four impellors in series (C)** Discharge from the first impellor is led into the suction of the second impellor, which then discharges into that of the third impellor, and so on to the fourth. Discharge pressure increases at each stage, although the quantity of water pumped remains the same as that pumped by a single impellor. While the quantity of water pumped is relatively small, the final discharge pressure is far greater than that achieved by modes A and B.

Whichever operational mode is used, pump output can be further adjusted by varying the pump's speed of rotation by raising or lowering the field rheostat. Before running the pump it was essential to have the rheostat fully wound back to the 'start' position to prevent power overload on start-up.

When discharging overboard, pump discharge pressure must be maintained at a higher value than relative sea pressure and is monitored on the series/parallel indicator, a pressure gauge with its face divided into three coloured segments indicating the range of discharge pressures appropriate to the

ABOVE Mono trim pump. (Plate 2, BR.1963/4)

RIGHT Trim pump operating station in the control room. (Author)

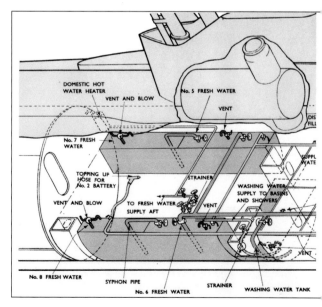

pump's three series/parallel settings (A, B or C above).

Maximum pump speeds when the boat is silent running are also indicated on the gauge. The degree of silence in operation is dependent on pump efficiency, which is variable with speed and depth. Whatever the given depth, the pump is run as indicated on the gauge, thereby ensuring that discharge is maintained above sea pressure.

Trim pump and system

If one considers a submarine to be lever balanced on a fulcrum at its centre, the purpose of the trim system is to balance it horizontally in the longitudinal plane. The hull can be tilted by moving the bodily weight of water afore or abaft by means of a pump centrally located in a pipe connecting trim tanks at either end of the submarine. The controlled operation of this system is called 'trimming the boat'.

Fitted internally, the two trim tanks are located amidships under the torpedo stowage compartments fore and aft, the tanks having capacities of 3,026 gallons (13,752 litres) and 2,203 gallons (10,013 litres) respectively. A trim line along the port side of the boat connects the tanks and the integral trim pump is located centrally in the control room. Both trim tanks have tank-side isolation valves and can be connected to the main line.

The trim pump operates on a principle related to ripple-wave reaction, similar to the movement of a rope fixed at one end when it is oscillated up and down, with the circumscribed sine curves thus created moving along its length. In a pump this movement is produced by means of a circular 'sine-wave'-shaped steel core (rotor) turning about its axis within a similarly shaped rubber tube (stator); the oscillating effect is produced by virtue of rotor and stator being offset. The direction of longitudinal movement of water through the pump depends on whether the rotor turns clockwise or anti-clockwise, as effected by moving the contact points between rotor and stator.

The actual trim pump comprises four pump units as described above working in parallel. Each of the four rotors is connected to a driveshaft, the four shafts being geared together and belt-driven from an electric motor. As the motor has a two-way starter, the pump can be reversed, so that the two pipe connections into the pump become either the suction or discharge. The pump unit has a discharge capacity of 12 tons (12,217 litres) per hour at atmospheric pressure and 7 tons (7,127.6 litres) per hour against a back pressure of 300psi; both pipes are fitted with relief valves set at 300psi. Pump efficiency is exceedingly low at low discharge pressures but very high when operating against a large head of water.

Distilling plant

The distilling plant converts seawater into fresh water. As it is somewhat limited in output, at 15 gallons per hour, priority in use of fresh

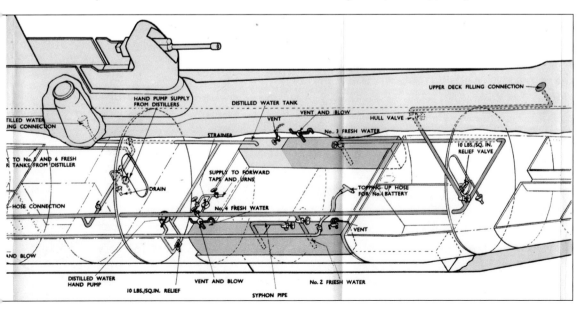

LEFT Fresh and distilled water systems. *(Plate 7/1, BR.4540/2507/2)*

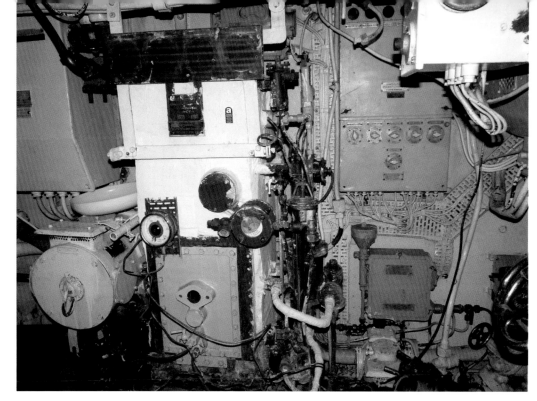

RIGHT Distilling plant evaporator with combined motor-driven distilled water, brine and chemical injection pump left, control panel right with salinometer below. *(Author)*

water is given to essential equipment and for cooking purposes. Because water availability for personal washing and laundering is very limited, the personal hygiene of the submarine crew is affected in service, such that the official motto of the Royal Navy Submarine Service, 'We Come Unseen', has often been parodied as 'We Come Unclean'.

The distilling plant carries out two main processes:
- Boiling of seawater and then distillation of the resulting condensate (vapour).
- Removal of the brine produced as a by-product of boiled seawater.

Entirely self-contained, the plant comprises a distiller body, vapour separator, immersion heater unit, distilled water and brine cooler/feed heater, feed tank evaporator, vapour compressor, a motor-driven pump, a distilled water control vessel, a salinometer (indicator gauge) and a control panel.

Seawater is boiled within an evaporator shell maintained at a vacuum, the heat source being provided by electrical heating elements within the shell. The reason for boiling under a vacuum is improved efficiency, as less energy is required. This vacuum is produced by means of a compressor that draws off the condensate.

Air-conditioning plant

Designed for operational service in warmer climates, particularly the Far East, the A-class boats were fitted with two air-conditioning plants providing dry, cooled air by means of a fan drawing air through an air cooler and passing it into the ventilation system ducting. Both plants, located below the control room, could be used either together or singularly. Each plant is designed to absorb 55,000 BTUs per hour from a volume of 36,000cu ft per hour. Each plant consists of a compressor, a condenser, a heat exchanger, a thermostatic gas control unit and cooling coils.

The compressor comprises a four-cylinder, single-stage, motor-driven machine taking its suction through the crankcase. In the crown of each piston is a suction valve that opens on the downward stroke, allowing gas to flow through the pistons into the cylinders. On the upward stroke the valves shut so that the gas is compressed and forced through spring-loaded discharge valves from where it passes over to the condenser.

The condenser is a standard straight-tube 'nest' set between two tube plates bolted to end covers, one of which has integrally cast cooling water inlets and outlets. Cooling water, supplied by the submarine's auxiliary circulating system, makes eight passes through the tube bank. The hot compressed gas passing into the top of the condenser will subsequently condense and fall to the bottom of the condenser and in so doing converts into a liquid, transferring its latent heat to the seawater used for cooling. The liquid refrigerant then passes through a heat exchanger before reaching the thermostatic control unit.

The heat exchanger is attached to the bottom of the air cooler and provides further cooling to the liquid refrigerant before passing it on to a regulator valve. The heat exchanger consists of a copper tube into which cold gas from the cooling coils passes back to the suction side of the compressor.

The thermostatic gas control unit regulates the flow of liquid refrigerant into the air cooler coils, thus controlling the temperature of the air leaving the cooler.

When the air-conditioning plant is in operation, liquid refrigerant from the condenser enters the thermostatic gas control unit via a strainer and then passes down through an internal tube on top of the regulator valve. When the gas leaving the air cooler is warm, it heats up the liquid refrigerant in a bulb, creating pressure in a sealed bellows. The increase in pressure expands the bellows, opening the regulator valve to admit liquid refrigerant to the cooling coils. By virtue of the fact that the pressure in the cooling coils is lower than the condenser, the liquid expands and evaporates. To change physical state, the liquid refrigerant needs latent heat that it absorbs together with 'sensible' (real) heat from the air flowing over the cooling coils. When the gas leaving the cooling coils is cold, it condenses the refrigerant in the bulb, thus reducing the pressure in the bellows; consequently the bellows will contract, allowing

BELOW Air-conditioning plant. *(Plate 23, BR.1963/4)*

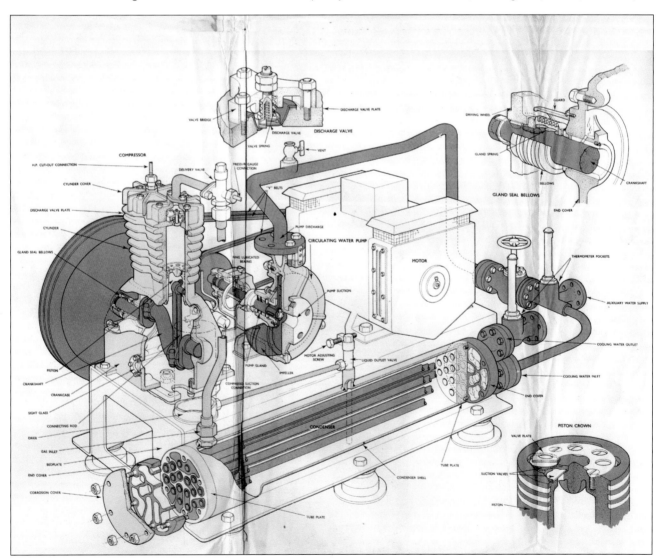

the regulator valve to close and restrict the flow of refrigerant to the cooler coils.

The refrigerant gas used in the plant was Freon 12 (also known as Arcton 6), which has the chemical name Dichlorodifluoromethane CCl_2F_2 and was chosen for the following properties:

- It is non-inflammable and odourless.
- It will liquefy and give up its latent heat at high temperatures.
- It will exist in both gas and liquid state at seawater temperature at reasonably low pressures, and has a freezing point well below that required for system operation.
- It is non-toxic, though it is vital that it does not leak into the submarine atmosphere in concentrated form.

Routine monitoring for system leaks was essential when a submarine was in service. Before the introduction of portable electronic detectors, this was undertaken using a halide, a small blowlamp that used alcohol fuel and a copper heating wire. The flame drew air through a flexible pipe attached to a small funnel that was held over a suspected leak; if refrigerant gas was leaking, the flame would burn blue-green in colour.

Freon escaping from closed systems into the atmosphere can prove dangerous. At very high temperatures (above 250°C) it will produce phosgene gas, which is similar to the mustard gas used during the First World War and produces the following effects in people exposed to it:

- Reddening and blistering of skin.
- If inhaled, blistering to the lining of the lungs, causing chronic impairment to breathing or even death.
- Deterioration of eye corneas, eventually rendering a victim blind.

Refrigerating system

The refrigerating system supplies a cold room, a cool room and a ready-use cupboard, and maintains them at their designed temperatures. The refrigeration plant, together with the cold room and cool room, are located with the auxiliary machine space, while the ready-use cupboard is located in the accommodation space above the store.

The 133cu ft cold room was the most refrigerated of the three facilities, with an operating temperature of 15–18°F (8.3–10°C), and was used for storage of meat and fish. The 35cu ft cool room ran at 30–35°F (16.6–19°C) and was used for diary products and other fats. The ready-use cupboard, with a capacity of only 12cu ft, was a 'convenience' that was used as required and kept at 35–40°F (19–22.2°C). The cold room and cool room are well insulated with cork and lined with zinc, all joints being soldered.

The system had a working capacity of 6,000 BTUs per hour and its main components were a motor-driven compressor, a condenser (with a circulating pump driven by the compressor), evaporator grids in the cooled compartments, and thermostatic expansion valves. As with the air-conditioning plant, the refrigerant gas used in the system was Freon 12, and detection of leaks was carried out in the same way.

Auxiliary air purification systems

Exhaled air contains carbon dioxide, a toxic gas that would accumulate in the submarine when submerged unless removed. Two simple methods of air purification can be made when submerged:

RIGHT Snort induction piping extending down from the hull, with ballast control gauges in the foreground. *(Author)*

FAR LEFT CO_2 absorption unit, accommodation section. *(Author)*

LEFT Oxygen generator, accommodation section. *(Author)*

- Raising the snort induction mast and replenishing air via the snort induction valve.
- Raising the snort induction mast together with the snort exhaust mast and running the diesels to recharge the batteries; all stale air will be consumed by the diesels in their combustion process.

Operational requirements and the need to avoid detection, of course, often meant that these simple methods were inappropriate, so two alternative methods of air purification were provided in the form of CO_2 absorption units and O_2 generators.

CO_2 absorption units

The function of the two CO_2 absorption units fitted to A-class submarines was to chemically remove carbon dioxide from the submarine's atmosphere in the event of an emergency and to deliver purified air with a reduced CO_2 content. These relatively compact machines are fitted in the seamen's mess near the fore escape compartment and under the plates at the main switchboard near the after escape compartment. Designed to operate with minimum vibration and noise levels, these units were virtually undetectable by an enemy submarine.

Extremely reliable in use, these units required little or no maintenance and are simple in operation. A motor-driven fan draws air contaminated with CO_2 down through four open soda-lime canisters. Purified air is then delivered into the submarine either through dedicated tubing or via the submarine's ventilation trunking. DC electrical power is supplied via a regulator that runs directly off the submarine's main battery, the regulator controlling the amount of purified air delivered into the submarine's ventilation trunking.

In the event that normal battery supply is unavailable, a changeover switch is provided to connect the unit to the submarine's seawater batteries. If this alternative power facility is also unavailable, the fan can also be driven by reduced air pressure, though it was recommended that caution be observed if using air in an escape situation (see Chapter 9). A special tool was provided for opening the soda-lime canisters.

O_2 generators

The purpose of the O_2 generators is to produce oxygen chemically by means of burning a candle comprising a compound of sodium chlorate and iron in powdered form. When ignited, the mixture smoulders at about 600°C and produces oxygen along with sodium chloride and iron oxide. The oxygen is released by thermal decomposition at a fixed rate equivalent to about 6.5 man hours of oxygen per kilogram of the mixture. A candle typically lasts for about an hour. Care must be taken that candles do not become contaminated with oil as this will cause them to explode rather than burn slowly.

The procedure for operating one of the O_2 generators is as follows:

1. Unscrew breech and lower it down.
2. Insert unburned candle.
3. Raise and shut the breech.
4. Lock the breech.
5. Unscrew the firing pin at the top of the unit.
6. Insert a 410 cartridge and screw firing pin back on.
7. Bang down on the firing pin with your hand – this action fires the cartridge and starts the candle.

The O_2 generator has two chambers on the side. The first chamber holds a filter to remove burned particles from the candle that then pass through a lower section filled with charcoal. The second chamber is where the filtered air passes upwards and out into the submarine's atmosphere through a tube inside which is a small ball – like a ping-pong ball – that bounces as air passes through, indicating audibly that the generator is running.

Fresh water and 'domestic' systems

Primarily used for 'domestic' needs, fresh water is contained in dedicated tanks that are initially filled from dockside facilities via a hose connection and hull valve. Besides supplying the hot water heater, the system also supplies 'non-domestic' services, such as topping up the main batteries. Once the boat is at sea, fresh water is replenished by means of the distilling plant, which converts seawater into fresh water. Tanks and capacities are listed in the accompanying table.

Crew washing facilities were initially limited to six washbasins and three showers dispersed between three individual bathrooms for officers (6 men, one washbasin), senior ratings (18 men, two washbasins) and junior ratings (65 men, three washbasins), each bathroom having one shower.

The provision of three showers was considered unnecessary and they were later removed and replaced with a single shower unit fitted inside one of the WC cubicles. This reduction was made for various reasons: to conserve limited fresh water (the distilling plant was not able to satisfy demand); *Alliance* later reverted to operating in cooler home waters where patrol times were much reduced, so there was less need for showers; and the crew were less concerned about showering when operating shorter patrols.

There are just four 'heads' (WC flushing lavatories) in the boat: one in the officers'

FRESH WATER SYSTEM TANKS AND CAPACITIES

Tank	Location	Capacity (gallons)
Number 1 fresh water tank	–	–
Number 2 fresh water tank	Starboard, outboard of number 1 battery tank	421
Number 3 fresh water tank	Port, outboard of number 1 battery tank	578
Number 4 fresh water tank	Starboard, outboard of number 2 battery tank	445
Number 5 fresh water tank	Port, outboard of number 2 battery tank	484
Number 6 fresh water tank	Starboard, outboard of number 2 battery tank	450
Number 7 fresh water tank	Port, outboard of number 2 battery tank	500
Number 8 fresh water tank	Starboard, aft of number 6 fresh water tank	318
Water washing tank	Starboard, forward of number 6 fresh water tank	626
Distilled water tank	Port, outboard of number 1 battery tank, aft of number 6 fresh water tank	315
Total capacity		**4,137**

Note The ship's book only gave total capacity for the eight fresh water tanks, at 3,812 gallons, so the individual tank capacities given above have been calculated from drawings.

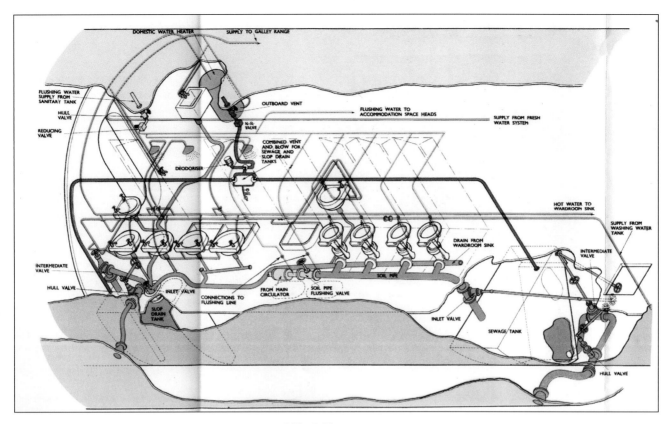

ABOVE Layout of bathrooms and WCs. (Plate 7/2, BR.4549/2507/2)

bathroom and three for the remainder of the crew, one of which was later converted into a shower (see above). A urinal was later fitted to compensate for the loss of one WC. When first in service, flushing water was supplied from a sanitary tank located within the conning tower. This tank was isolated at the system hull valve when going deep. It was later removed and flushing water was supplied from the main circulating system via a reducer.

Waste water from washbasins, shower units and the sinks in the galley and wardroom pantry drain away into the slop-drain sewage tank, fitted on the starboard side aft of number 8 fresh water tank. Flushing water and waste from all lavatories passes to a common soil pipe leading into the main sewage tank of 168 gallons (763.6 litres) located on the starboard side forward of number 8 fresh water tank. Both slop-drain and sewage tanks can be flushed through with seawater supplied from the main circulating system, and both are provided with a combined vent/blow facility using high-pressure air to discharge their contents overboard. Although discharging of these tanks was normally undertaken at night, operational requirements sometimes made it inappropriate to do so.

Oily bilge system

The function of the oily bilge system is to remove unnecessary drainage water and oil, by means of low-pressure bilge pumps, from various tanks and compartments and discharge it overboard. The tanks and compartments are as follows:
- Fore ends, torpedo tube space bilge.
- Fore torpedo operating tank.
- Forward torpedo stowage compartment.
- Number 1 battery tank.
- Auxiliary machinery space.
- Both periscope wells.
- Radar mast well.
- Air-conditioning space bilge.
- Gun tower drain.
- Conning tower drain.
- Number 2 battery tank.
- Engine room bilge.
- Motor room bilge.
- Oily bilge tank.
- After torpedo stowage compartment.
- After torpedo operating tank.
- After ends bilge.

Oily bilge water is only discharged when operational conditions allow; when running deep the tank can be discharged by means of the ballast pump.

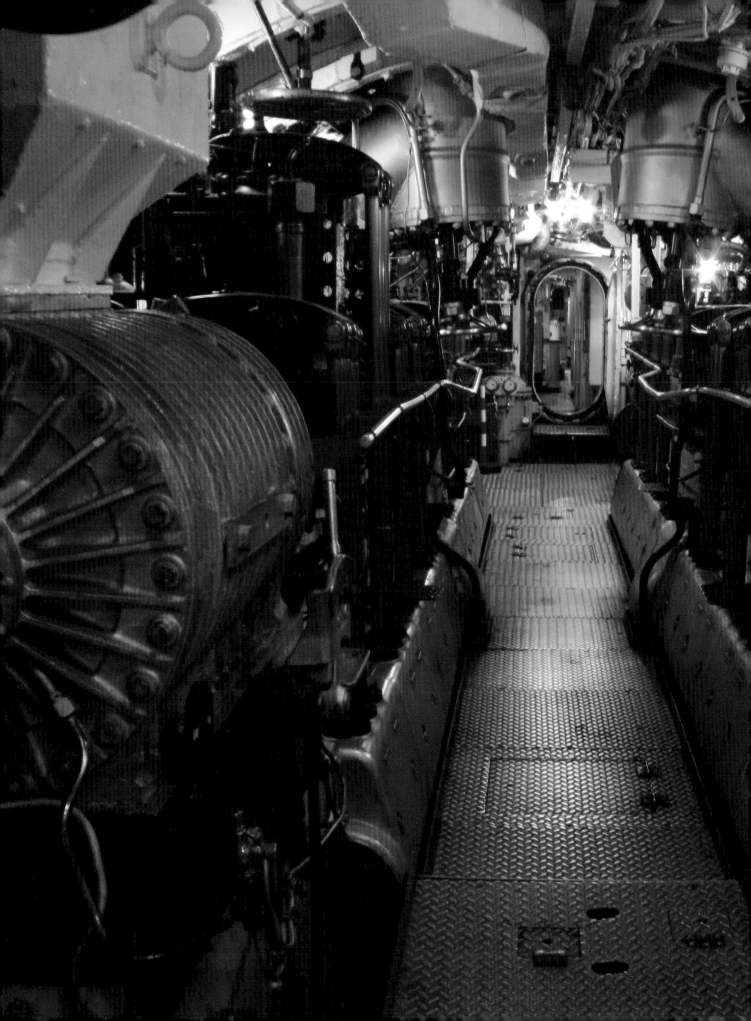

Chapter Four

Alliance's propulsion

HMS *Alliance* was built for speed and endurance. Her two 2,150bhp Vickers diesel engines could propel her to 18½kts on the surface, but when slowed to 11kts she could go 10,500 miles without refuelling. In trials in 1947 she spent 30 days continuously submerged and covered 3,193 miles.

OPPOSITE Engine room looking aft to bulkhead 126 W/T, starboard main engine (left), port engine (right), group exhaust valves are seen halfway along.
(Jonathan Falconer)

ABOVE Engine room looking aft, 1978.

BELOW Vickers main engine general arrangement. *(Plate 1, BR.1953/9A)*

Diesel main engines

The two main engines are eight-cylinder, single-acting, four-stroke supercharged Vickers diesels driving separate shafts and each producing power output of 2,150bhp at 4,360rpm, providing the boat with a surface speed of 18½kts. Each engine was tested to 10 per cent overload for two hours – once after manufacture and repeatedly following respective maintenance. A secondary function was to produce DC electrical power to charge the submarine's main battery, which supplied all electrically driven equipment.

Engine frame

This consists of ten separate columns bolted to the engine bedplate with the cylinder heads forming an integral girder across the top of the columns. The bedplate, made from iron plates, is welded to ten cast-steel bedplate girders. Housings for the crankshaft main bearings are machined in the bedplate girders. The columns each comprise two drop-forged legs with mild-steel tie plates between them, forming an A-shaped frame. Cylinder heads are furnished with cast flanges machined on their forward

In *Alliance* the propulsion system consists of a direct mechanical connection between the engine and propeller, switching between diesel engines for surface running and electric motors powered from the main batteries when submerged. In effect this is a 'parallel' type of hybrid system, since the motor and engine are coupled to the same shaft, with the option to disconnect the engine for submerged operation, and using the motor as a generator to recharge the batteries while on the surface.

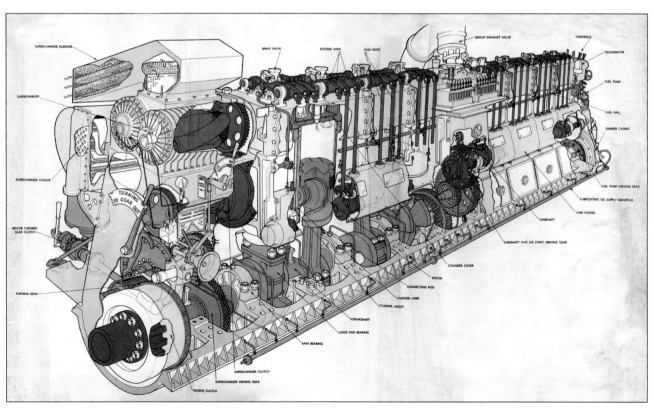

and after ends through which they are rigidly bolted and fitted together. A centrepiece fitted between number 4 and number 5 cylinder heads encloses the centre gear bay with special end pieces to complete the girder.

Crankcase

Enclosed by side doors, a diaphragm plate and tank tops underneath, the crankcase is completed by end and side closure plates. An engine bearer welded to the crankcase closure plate is machined to agree with propeller shaft alignment. Once positioned, the engine bearer is held down with 36 studs and 44 bolts. Eight clearance holes are provided to receive jacking bolts for aligning requirements.

Crankshaft

The crankshaft is built up of two forged sections of oil-toughened carbon, with each section bored throughout to a diameter of 6in. Each component, a mirror image of its counterpart with respect to the crank throws, is coupled together at the centre with spigoted flanges and 12 fitted bolts. These flanges also secure the split driving wheels for the engine-driven lubricating oil pump and camshaft. Thrust faces are machined on crank webs either side of the centre coupling.

Each section of the crank is flanged at its end, the foremost flange carrying the split fuel pump drive hub and the inner part of the vibration damper. The aftermost flange carries the forward star plate of the engine clutch secured with 12 bolts. Two holes are drilled

MAIN ENGINE STATISTICS AND SPECIFICATIONS

Total engine weight including supercharger and turning gear	35.5 tons
Total engine weight 'wet'	38.0 tons
Weight of clutch	1.5 tons
Maximum cylinder pressure at maximum bhp	690psi
Maximum exhaust temperature at maximum bhp	760°F
Maximum exhaust temperature at maximum bhp (supercharged)	850°F
Supercharger induction air pressure to full power	4.72psi
Induction air temperature at full power	150°F
Supercharger speed at full power	1,865rpm
Power required to drive supercharger at full power	200bhp
Normal air start pressure	350psi
Air start bottle pressure	900psi
Cylinder firing order, port engine	1-6-7-4-8-3-2-5
Cylinder firing order, starboard engine	1-5-2-3-8-4-7-6
Bore	17.4in
Stroke	18.5in
Compression ratio	12:65

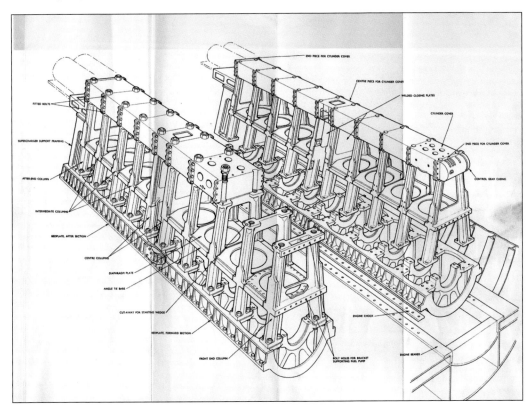

LEFT Construction of engine frame.
(Plate 2, BR.1963/9A)

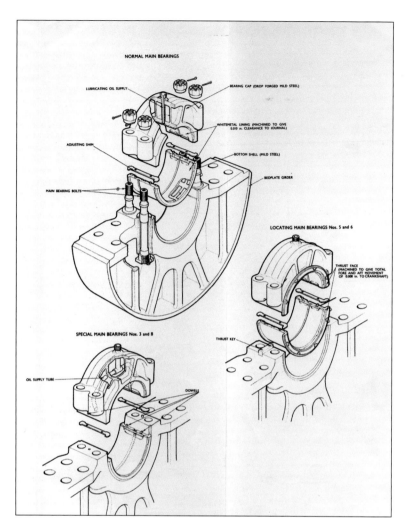

ABOVE Main bearings. *(Plate 4, BR.1963/9A)*

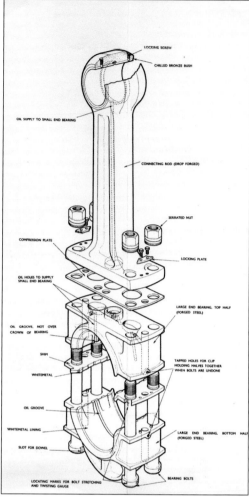

ABOVE RIGHT Con rod assembly. *(Plate 5, BR.1963/9A)*

through the after flange to allow the 12 star-plate bolts to be withdrawn.

Balance weights are machined upon the shaft on numbers 1, 4, 5 and 8 webs. Lubrication for each of the large end bearings and the clutch sleeve is provided through a brass tube expanded into the shaft at the centre of all main bearing journals except numbers 3 and 8, which require greater oil supply due to the extra load on the bearing created by adjacent cylinders firing consecutively. The clutch star plate is lubricated from the clutch sleeve bearing through a hole passing from the centre of this bearing through the after flange.

A Vickers-Sandner torsional vibration damper is fitted to the forward end of the crankshaft.

Bearings

The engine has 11 main bearings for the crankshaft. Numbers 5 and 6, designated as locating bearings, have white-metal thrust faces machined upon them; numbers 3 and 8 are special bearings while the remaining seven are normal. Lower half-bearing housings are machined into the engine girders and white-metal-lined, mild-steel shells are fitted in these housings. No shells are fitted in the top half, the bearing cap being lined with white metal.

Connecting rod assembly

The connecting rod assembly comprises two main parts: the connecting rod itself with small-end bush and the large-end bearing. The con rod is a mild-steel drop forging of H section with an eye at the top for the small-end bush and a foot at the bottom to receive the large-end bearing. Each rod is bored through vertically to convey lubricating oil. The small-end bush is of chilled bronze and pressed into the eye of the con rod to form a bearing for the gudgeon pin and secured with locking screws through the top of the rod. The large-end bearing is manufactured in two halves of drop-forged steel, lined with Hoyts 11A gunmetal and fashioned with oil grooves in the horns. Dowel pins are fitted in the butts of the bottom half for locating needs.

Pistons and rings

Made from cast aluminium, the pistons each have a toroidal combustion space formed in the crown. Each piston is machined with four valve recesses in the top to clear the valves, two of which are tapped to take the lifting gear. Integrally cast bosses are bored horizontally to form bearings for the gudgeon pin. The periphery at the top and skirt of the piston is machined with grooves to receive the piston rings.

The piston rings comprise six cast-iron impulse rings and one scraper ring. The former are of rectangular cross-section with a radius on their outer edges and a chamfer on the inner edge. The scraper ring with sharp corners is chamfered on its lower edge. The gudgeon pins comprise case-hardened and ground hollow steel pins.

Superchargers

There are two superchargers, one fitted to each engine. Each supercharger is a positive-displacement, three-lobe Rootes type driven off the after end of the camshaft through a train of single helical gears and a friction plate, with a cushion drive incorporated within the train.

Centre drive

This consists of a train of single helical wheels driven from the centre of the crankshaft drives for the camshaft, the air start distributor and mechanical lubricator. A second train of single helical gearing, also bolted at the centre of the engine, drives the integral lubricating pump.

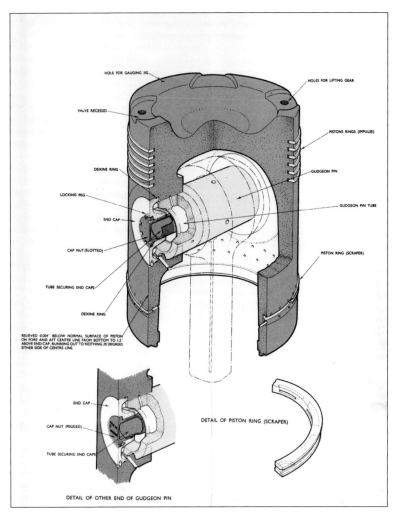

Fuel injection

Fuel is supplied to a common manifold rail by two blocks of three pumps driven off the forward end of the camshaft through a cushion driving wheel set in a train of five wheels incorporating drives to a fuel booster pump and an Amal drain pump. The fuel spray valves are mechanically operated, their timing being independent of the pumps.

ABOVE Piston assembly. *(Plate 6, BR.1963/9A)*

FAR LEFT Port main engine supercharger with cooler overhead. *(Author)*

LEFT Port engine cylinder head fuel spray valve injector assembly, inlet and exhaust valve rocker arms and exhaust pipe to exhaust manifold. *(Author)*

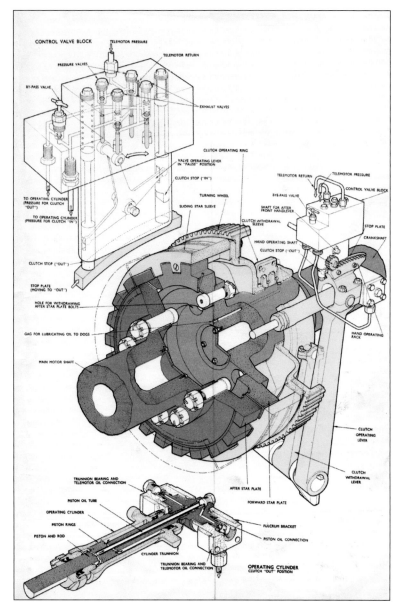

LEFT Engine clutch operating gear. *(Plate 14, BR.1963/9A)*

ABOVE Port main engine, main clutch lever (centre) and turning gear clutch (bottom left). *(Author)*

Engine clutch

The main drive of the engine is transmitted through a telemotor-operated (hydraulic) dog clutch connected to the after end of the crankshaft.

Main engine lubricating oil system

The main lubricating oil system for the diesel engines comprises three tanks located under

BELOW LEFT Main engine diesel fuel oil separator port side abaft HP air compressor. *(Author)*

BELOW Main engine lubricating oil filter unit. *(Author)*

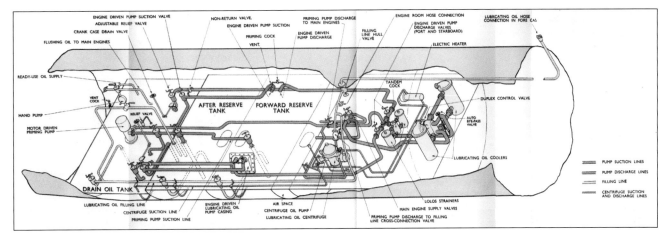

ABOVE Lubricating oil system.
(Plate 20, BR.1963/9A)

the engine/motor room: these are the forward and after reserve tanks of 1,164 gallons (5,290.4 litres) in each and the drain oil tank (1,132 gallons, 5,145.9 litres). The lubricating oil used was OMD 112, the naval category OMD indicating 'oil mineral detergent' and the number relating to viscosity.

Driven by the main motor, the lubricating oil pump takes its suction from whichever lubricating oil tank is selected via the suction stop valves and discharges oil through one of a set of three Lolos strainers (two were off line). Oil then passes to one of a set of two oil coolers (one was off line) and thence to the engine-driven lubricating oil pump, from which it is distributed to bearings and associated integral lubricated components. A lubricating oil centrifugal separator (see below) is incorporated into the system to purify the oil and remove residue (water, carbon and material contaminates), the oil being preheated beforehand. The centrifugal separator can be indirectly put on line to any of the three lubricating oil tanks to maintain purity.

Oil was to be tested daily with silver nitrate to identify seawater cross-contamination at the coolers.

Circulating system

Three 4½in (114.3mm) Drysdale motor-driven seawater pumps are fitted port, starboard and mid-line in the engine room, each supplying 18,000 gallons (81,810 litres) per hour at 30psi (2.06 bar). The mechanical component

ABOVE Starboard main engine lubricating oil system distribution manifold. *(Author)*

BELOW Starboard group exhaust valve. *(Author)*

89

comprises a Monel metal impellor keyed to a manganese-bronze shaft. Each pump has its own sea inlet, hull valve and weed trap. Sharing a common line, the system also permits all three pumps to take suction from either inlet; pump discharges can also be cross-connected. Under normal conditions pumps run in units, with port and starboard pumps supplying their respective engines and lubricating oil coolers, the discharge from the engine passing to the muffler tank. System test pressures were as follows: suction pipes, 300psi (20.7 bar); discharge to engines, 60psi (4.4 bar); and the after-services system, 300psi (20.7 bar).

The mid-line pump supplies the after-services system with seawater at 300psi to the group exhaust valves, supercharger coolers, main motor coolers, main motor bearings, thrust block and stern glands. The return side is led to a common discharge pipe and discharges via a common overboard discharge pipe and hull valve, and supplements the muffler tank cooling spray. The outlet from the group exhaust valve passes to the external exhaust pipe hull valve and continues to supply the exhaust pipe in the muffler tank, where its final discharge acts as the muffler valve spray.

Running the main engines

The following operational procedures are taken from BR.1963 (9A) Handbook for A Class Submarines Part 9A Vickers Main Engine. Main engines sometimes had to be run when alongside the jetty to charge main batteries before proceeding to sea.

Prerequisites

The following prerequisites are necessary:
1. All lubricating oil, fuel oil and clutch telemotor systems are operable.
2. Circulating cooling water system is operable.
3. Fuel/seawater compensating water system is open to fuel tanks.
4. Engines have been turned over by hand to check free to turn.
5. Engine controls are fully operational.
6. Fuel oil centrifuge is operable.
7. Lubricating oil centrifuge is operable.
8. Main propeller shaft brake is off and shafting is free to turn.
9. Associated shaft plummer blocks and bearings are operable.
10. Shaft stern tube glands are operable with seawater supply system open.
11. Main motor is operable.
12. Air start bottle group is fully charged to 900psi.
13. Both high-pressure air compressors are operable.
14. Communications between control room and engine room are established.

The following quotation taken from document BR. 1963 9A (see Appendices) is noteworthy: 'Detailed preparations for starting and running an engine are not described, for these belong essentially to the Engineer Officer's Standing Orders of individual submarines, but some simple reminders are given.'

Before starting main engines

1. Check the position of the engine, tail and supercharger clutches.
2. Check the oil level in all shaft bearings.
3. Start the lubricating oil centrifuge and test the drain oil tank for water.
4. Prime the fuel system and vent the spray valve.
5. Start the motor-driven oil pump and circulators and vent the systems.
6. Turn the engine two revolutions with the turning gear.
7. Take out the turning gear and blow round, sighting all cocks and drains on the engine induction main, the exhaust trunking and supercharger coolers.

Starting main engines

Using control levers, start main engine. Once engine has started, check round for any oil and/or fuel leaks. **Note:** if any leaks are found (especially fuel-related) or potentially hazardous, which cannot be simply and practically rectified, the engine is to be shut down to make good necessary repairs. Commence monitoring running parameters ensuring the following:
1. Maintain the cylinder head water outlet temperature at 120°F (66.6°C) and the engine inlet temperature at 90°F (50°C), using the looping system.
2. Ensure adequate water supply to the muffler tank. If the supply to the main motors are

open this will be sufficient but if they are throttled, the cross-connection on the after services must be opened and adjusted to give the necessary circulation to the thrust blocks and muffler tank.

3 Any of the three circulating pumps can be used for all services at reduced powers.
4 One oil cooler can be isolated, in the event of failure, on both the oil and water sides, in which case the engines can be run to the limit of capacity of the other cooler to keep the lubricating oil supply at 105°F (58°C).
5 Run the oil fuel centrifuge at all times of topping up gravity tank and clean it daily. Water present in the fuel will pass to the main drain tank, which should be pumped out to the bilge when the quantity becomes excessive.
6 Maintain lubricating oil at 170°F (94°C) by use of the heater in the centrifuge circuit.
7 Oil supply to the engine must never be throttled, but the pressure regulated on the adjustable relief valve fitted on the engine-driven pump discharge.
8 Set mechanical lubricators to four drops per minute, except in the case of a newly fitted liner, when they are increased to ten drops per minute on the individual pumps supplying that liner.
9 To avoid flooding the lubricating oil centrifuge, the vent on the hand pump suction line should be open.
10 Check the exhaust temperatures for irregularities in combustion and mechanical failure in the valve gear.
11 All drains should be opened frequently.

Shutting down main engines

Ignoring the need to stop the engine(s) as routinely ordered, they are to be immediately shut down in the event of:
1 Bow-down angle exceeds six degrees for longer than two minutes as this will prevent lubricating oil draining back to the drain oil tank.
2 Running parameters cannot be maintained to permit safe operation.
3 Stop snorting.
4 Diving the boat.
5 Emergency crash dive.
6 Fire (in the boat).

Notes: Item 1 affects lubrication; items 3 to 6 will necessitate rapid action to be taken, group exhaust and muffler valves being instantly shut.

Maintenance of the main engines

To enable work to be undertaken, a total of 127 special tools were supplied for maintaining the Vickers diesel main engines. Those working on maintenance and repair had to be highly skilled in the practice of grinding valve seats and setting up of clearances. The following maintenance procedures are taken from BR.1963 (9A) Handbook for A Class Submarines Part 9A Vickers Main Engine.

LEFT Port main engine controls. *(Author)*

BELOW *Alliance* crew members in the engine room, 1953.

Routine running maintenance

Routine maintenance procedures were as follows:
- Clean lubricating oil separator twice daily.
- Clean fuel oil separator once daily.
- Clean lubricating oil filters as required.

General running repairs

The most common repairs expected to be undertaken at sea were as follows:
- **Pistons** Replace pistons, change impulse and scraper rings and gudgeon pins. Also disengaging a complete piston and connecting rod if damaged (may be undertaken under extreme circumstances providing that fuel is fully isolated from the affected cylinder and that disruption to firing order is accepted).
- **Induction and exhaust valves** Replace or reset valves and/or replace springs.
- **Connecting rods** Replace small-end bearing bush and big-end bearing linings.
- **Fuel system** Replace fuel injectors, reset spray valve levers and spray valve tappets, and recalibrate fuel injector spray valves (this would preferably need test shop facilities for calibration).
- **Air start system** Maintain/reset or recalibrate air start valve (this would preferably need test shop facilities for calibration).
- **Miscellaneous** Adjust or replace lever pushrod pins; replace crankshaft bearings; replace cylinder liners; clean lubricating oil filters.

Dockyard refit and maintenance work

Besides the work listed above, general work (excluding major damage or component failure) undertaken in the dockyard would comprise the following:

- **Main engine** Inspect and/or change all camshaft bearings, crankshaft bearings, camshafts and respective pushrods, and all thrust plates.
- **Pumps** Overhaul cooling water pumps, lubricating pumps and fuel pumps.
- **Clutch** Inspect and/or change clutch drive forward and after star plates, sliding star sleeve, operating ring, turning wheel and withdrawal sleeve.
- **Supercharger** Inspect and/or overhaul supercharger drive and respective components, supercharger clutch operating gear and supercharger induction valve.
- **Miscellaneous** Inspect and/or overhaul air start motor, all associated coolers, and pressure test all related systems.

Main electric motors and associated components

The main shafting between the main engine clutch and the propeller comprises four lengths supported by the main motor bearings: the thrust block, the inner and outer stern tube bushes, and the 'A' bracket at its extremity. The end section between the propeller and the first coupling, located immediately forward of the inner stern tube bush, is called the tail shaft. This can be fully withdrawn by retarding it outboard.

Couplings for the remaining three lengths of shafting – intermediate shaft, thrust shaft and main motor shaft – are situated at the tail shaft brake and tail clutch respectively. All sections of shafting are hollow, the open ends being sealed by a screwed plug locked by a countersunk screw. All fixed shaft couplings have spigoted joints and are fastened together

RIGHT Engine room: underside of diesel main engine cylinder head showing inlet and exhaust valves. *(Jonathan Falconer)*

FAR RIGHT Engine room: diesel main engine fuel injector and cylinder head valve rocker arms. *(Jonathan Falconer)*

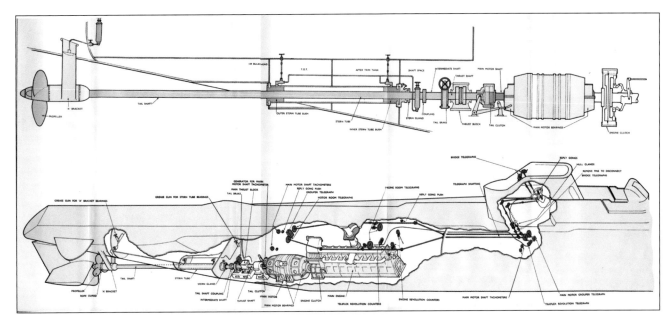

ABOVE Propeller shafting and telegraphs. *(Plate 10, BR.1963/8)*

at their flanges with six bolts equidistantly spaced. The diameter of the shafting up to the main motor is 7in (177mm) apart from two sections of increased diameter: the area of the 'A'-bracket bearing and the inner/outer stern tube bushes, where diameter is 7.5in (190mm); and around the tail clutch, where diameter is 7.25in (184mm).

Discussing one propulsion shaft unit only, commencing from the after end of each main engine, the main shafting comprises 16 in-line components as follows:

Main engine clutch

This disengages the diesel main engine from the main shafting in order to revert into battery drive using the main motor as the motive propulsion drive once the boat has dived. In brief, this consists of two star plates, one bolted to the main engine crankshaft, the other on the main motor drive shaft. The clutch is operated by hydraulic power or by hand by means of a sliding sleeve and operating ring. The assembly is contained within the casing that forms part of the engine crankcase. Power operation is controlled by one of two hand levers, one at the engine controls the other next to the clutch. Hand operation is provided by a hand wheel near the clutch. Interlocks are fitted between power controls and main motor switches. Major components comprise star plates, a sliding star sleeve, a clutch-operating ring, a control valve block and an operating cylinder.

The two clutch-operating levers are connected to a valve-operating lever that moves four tappet valves in the control valve block and in so doing connect telemotor (hydraulic) pressure or return to either side of the piston in the telemotor operating cylinder. Operating in power control, the engine clutch control levers have five positions:

- Mid – all valves shut; stops raised.
- Pause to out – pressure and return valves to clutch out fully open; both stops raised.
- Clutch out – all valves shut, out stop lowered.
- Pause to in – pressure and return valves to clutch in fully open; both stops raised.
- Clutch in – all valves shut, in stop lowered.

Movement of the levers to either of the 'pause' positions admits telemotor pressure to, and return from, the respective side of the piston in the operating cylinder.

When operating by hand control, a pinion mounted on the hand wheel engages with a

LEFT Starboard main engine turning gear clutch used to turn over the crankshaft before starting the engine. *(Author)*

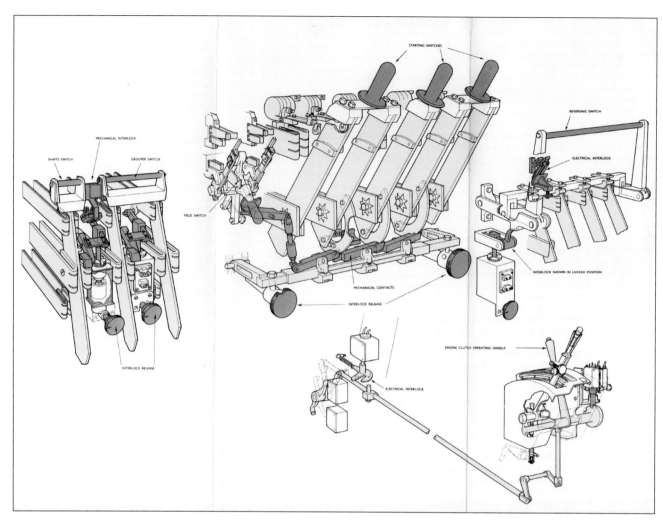

ABOVE A-class switchboard – electrical and mechanical interlocks. *(Plate 10, BR.1963/10)*

rack bolted with the stop plate on the operating lever and transmits to the clutch. Before operating the hand wheel, the following two points must be observed:

Control levers are put in the desired direction, ensuring that the correct stop is raised.

The telemotor (hydraulic) bypass valve must be opened on the control valve block.

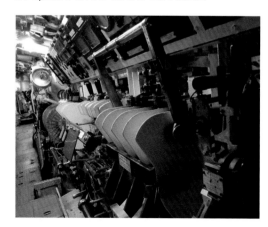

RIGHT Starboard main motor switchboard with knife switch gear and field regulator. Note the main motor telegraph towards the end. *(Author)*

Main motor

Fitted indirectly abaft the main engine clutch with its own integral motor shaft, each main motor comprises a compound wound armature encompassing a field magnet with two windings, and a shunt coil and series coil contained within a watertight casing. Two 500-watt electrical heating elements are fitted within the casing to prevent the motor getting too cold. To stop the motor overheating a variable-speed fan is fitted to the outer casing; drawing in air from the motor room and discharging it into the motor casing, the heated air passes through seawater coolers before exhausting into the motor room.

Main motor shaft and bearings

The main motor shaft forms the armature of the main motor discussed above. Two main motor bearings are fitted per shaft, one afore and one abaft each main motor. The foremost bearing

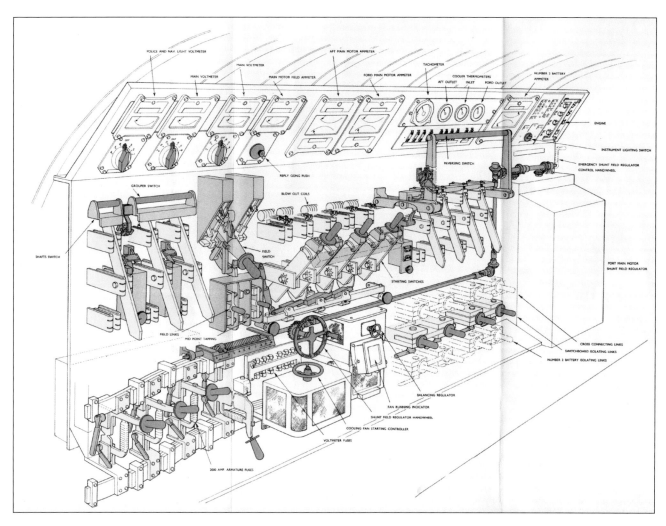

ABOVE A-class switchboard, general layout, port side. *(Plate 8, BR.1963/10)*

serves as a journal bearing while the aftermost bearing is a combined journal and thrust bearing. Contained within the same casing, this bearing also forms the drive for the electrical tachometer (revolution indicator). Both the fore and after main motor shaft bearings serve the same function as the plummer blocks used in surface vessels which, having much longer prop shafts, need a greater number of alignment supports along their length.

Tail clutch

This hand-operated clutch disengages the main motor shaft from the tail shaft. It can only be engaged when the clutch dogs of the two star plates, which are connected to the opposing collar faces on the two lengths of shaft, are directly synchronised in alignment connecting the two lengths of shaft. When disengaged the foremost plate is free to rotate independently of the after plate and sleeve. Each clutch is furnished with a long operating handle fitted on the inboard side of each clutch. The sliding-sleeve assembly comprises three separate components: the clutch-operating sleeve, the clutch-operating ring and the sliding-star sleeve.

Thrust shaft

This is a short length of shafting passing through the thrust block. Its main component, the

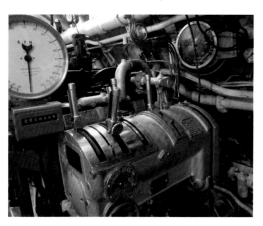

LEFT Port main engine controls with tachometer (revolution indicator) to the left and order telegraph (right). *(Author)*

thrust collar, is radially fitted on both sides with heavy steel horseshoe pieces that mechanically interact with the thrust block thrust pads.

Thrust block

This transmits propeller thrust to the hull body. It comprises a cast-steel lower-half casing and a light-alloy upper-half casing, the two enclosing the integral thrust collar of the thrust shaft. Both casings are bolted together at their flanges with 12 steel bolts. The upper casing serves only to prevent oil splash. The lower casing, which fully takes the workload, is rigidly fastened by 16 holding-down bolts to webbed longitudinal seatings that are welded to the hull frames.

Six segmental-shaped, gunmetal thrust pads are radially fitted opposite each side of the thrust shaft collar and each thrust pad is lined with white metal and shaped so that it can rock on its axis. It is this facility that forms the principal motion of permitting oil under pressure to form a wedge between the thrust pads and the thrust shaft collar. The correct oil level in the thrust block sump is 5½in (14cm).

Intermediate shaft

The same diameter as the rest of the shafting, the intermediate shaft provides a facility to disengage sections of shafting for the following maintenance/refitting reasons:
- Complete removal of the tail shaft.
- Removal of the main motor shaft.
- Removal of the thrust shaft.

The intermediate shaft is formed with two flanges. The foremost flange of the intermediate shaft serves to unite with the after flange of the thrust shaft, and the shaft brake is fitted at this union. The aftermost flange of the intermediate shaft unites with the fore flange of the tail shaft, this point being the tail shaft loose coupling.

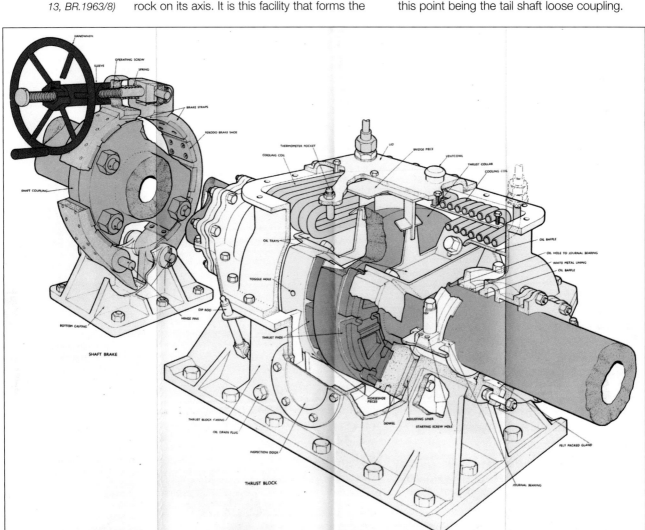

BELOW Shaft brake and thrust block. *(Plate 13, BR.1963/8)*

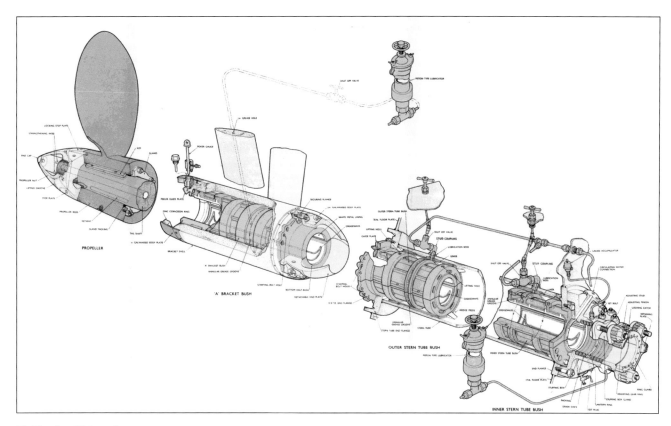

ABOVE **Tail shaft bearings, stern gland and propeller.** *(Plate 12, BR.1963/8)*

Tail shaft brake

This comprises two Ferodo-lined steel straps hinged to a bottom casting, the two straps being drawn together by a hand-wheel operating screw. The primary purpose of the tail shaft brake is to halt the rotation of a trailing shaft at slow speeds and, providing that the shoes bear evenly upon the shaft coupling, it will properly function at speeds up to 260rpm. Operating at speeds higher than this limit causes friction damage to both coupling and shoes.

Tail shaft loose coupling

This is the flanged union between the intermediate shaft and the foremost flange of the tail shaft, the disengagement of which permits the entire tail shaft to be withdrawn or disengaged from the respective motive propulsion units. The after flange of the intermediate shaft and fore flange of the tail shaft are aligned, married together and bolted with six bolts that are radially positioned and equidistantly spaced.

Tail shaft

The longest length of shafting, the tail shaft extends through the stern tube to beyond the 'A' bracket, its extremity being tapered to receive the propeller boss.

Stern gland

This consists of a stuffing box containing seven turns of ¾in (19mm) square-section Alanite packing with a 'lantern ring' fitted between the third and fourth turns. The whole assembly has a stuffing box gland, a gland-adjusting ring, a ring guard and four adjusting pinions. Excluding the packing rings and pinions, all parts are made in two halves to enable removal or assembly with the shaft in place. The entire assembly is connected to the circulating seawater system for lubrication.

Stern tube

Acting as a guide for the tail shaft, this fabricated steel component was welded into position when the pressure hull had been completed. Made with a wall thickness of ¾in (19mm) and internal diameter of 14½in (355mm), it is fitted with liners and lubrication bosses. Once fitted, it was pressure-tested to 300psi (20.7 bar) and thus was safe at a depth of 600ft (182.88m).

ABOVE Starboard propeller shaft, 'A' bracket and restored propellers. Note close proximity of after hydroplane and rudder to take good effect of propeller wash. *(Author)*

Stern tube bushes

The bush fitted at the inner end of the stern tube provided the third bearing from aft to support the propeller shaft. The bush fitted at the after end of the stern tube provided a bearing for the shaft; the end of its steel liner has an internal diameter of 12.875in (32.7cm) along a length of 3ft (91.4cm).

'A' bracket

Consisting of a forged steel shell, this is secured by means of two forged steel arms welded to the plating of number 5 main ballast tank. The bracket has an internal bush made in two halves of cast steel with a white-metal lining; grease ways are incorporated for lubrication.

Propeller

Made from manganese-bronze, each propeller has three blades, the port propeller being left-handed, the starboard right-handed to maintain equal thrust (ahead or astern) and prevent bias of veering to port or starboard. Each propeller is secured on to its shaft by a manganese-bronze, cone-shaped propeller nut that forces the propeller boss on to the tapered end of the shaft, the internal bore of the boss being tapered to match the shaft. The propeller boss is located on to the shaft by two keys that are fitted into diametrically opposite keyways. Each key is round-ended in the shaft keyway and square-ended in the boss keyway, the keys being fastened with cheese-headed screws.

Propeller maintenance

The following procedure for maintenance of propellers is taken from Naval Engineering Practice Vols 1 & 2 (HMSO, 1959) and personal practical experience.

Propellers were to be inspected six-monthly for the following points:

- Tightness and security of the propeller nut and gland; screws holding propeller nut stop plate to be checked for tightness; the eddy plate between 'A' bracket and propeller to be removed to examine prop gland. Use this opportunity to inspect zinc corrosion of ring anode.
- Examine propeller tip blade edges for 'fining' damage as this may cause the propeller to 'sing'. Excessive propeller noise could lead to detection by an enemy.
- Examine blade for erosion, which can have a serious effect on blade performance. Small air bubbles that form in the rough surfaces could eventually explode with force and increase drag on the surfaces, reducing design thrust. This can be effectively reduced by cleaning and polishing the blades to give as smooth a finish as possible, using wooden scrapers, fine emery paper and soft wire brushes.
- Examine security and condition of the integral balance pieces if fitted; often of a dissimilar metal, these could work loose and corrode.

Snort induction and exhaust systems

Introduced to British submarines towards the end of the Second World War, snort induction and exhaust systems provided a submarine with the ability to virtually stay fully submerged while running the diesel main engines to propel the boat and at the same time generating electricity to recharge the batteries. An explanation of the evolutionary development of this system in A-class submarines is given in Chapter 2.

When originally fitted into the A-class boats, this feature comprised a combined snort induction and exhaust mast, the two being attached to each other. Housed within the casing at the after end of the conning tower on the port side, the head of the induction mast was originally fitted with a ball-type float valve, which automatically shut off the mast tube from seawater ingress when the boat inadvertently submerged or was temporarily covered by heavy seas. Evidently keeping the boat level and well trimmed in the longitudinal plane is essential when operating at snort depth. The 1949 snort system with its ring-type float head valve comprised the following components.

Ring float-controlled head valve

This assembly was fitted within a galvanised tube guide pipe bolted to the snort mast, the float consisting of four components constructed of aluminium alloy and alloy sheet. To prevent the valve freezing in arctic conditions, a heater element was fitted within the space between the head valve and the hood, the electrical supply running up the inside of the guide pipe and induction tube via a gland located at the foot of the snort mast. The hood covering the entire assembly was made of mild steel and the bottom half of it was pierced by ten mesh-covered flood and drain holes.

Induction tube

This mast was formed of a galvanised steel streamlined tube 28ft (8.5m) long, the whole being strengthened by an internal perforated plate. The head valve and guide pipe were attached by a flange secured with 16 bolts. The base, referred to as a trunnion bearing, comprises a machined steel forging about which the mast pivoted. When fully raised, the mast is secured to the after end of the conning tower by a locking plate and bolt secured from within the submarine.

Water trap

Comprising a mild-steel cylinder with integral separator baffle plate, the water trap prevented any seawater ingress that may occur when 'snorting' (running submerged with the diesels in operation). All such water drained directly down into the 'R' port compensating tank via a drain valve, operationally referred to as 'snort drain one', which was opened prior to snorting and remained open for the duration of the snorting operation.

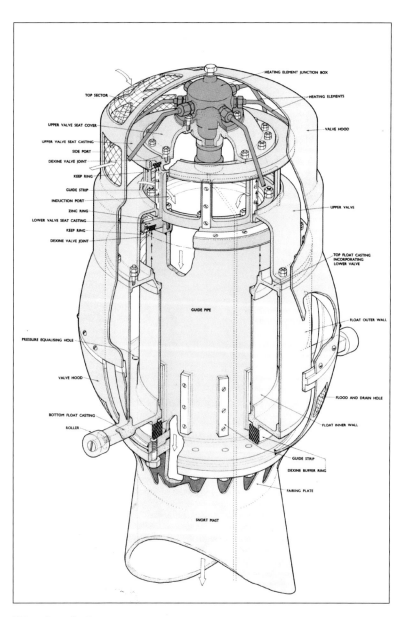

ABOVE Ring float-controlled head valve. *(Plate 6, BR.1963/8)*

Mast raising ram

The induction mast was raised and lowered by means of a hydraulic cylinder providing leverage to a forked, cranked arm welded to the base of the mast, hydraulic pressure being supplied from the submarine's main telemotor system. To avoid oil leakage betraying the submarine's position, the telemotor cylinder and associated piping were hermetically contained within a pressure-tight steel covering.

Snort induction hull valve

This is located within a cast-steel, pressure-tight dome fitted to a specially shaped pad bolted to the pressure hull above the engine room. The valve was opened and shut by means of

RIGHT Telemotor-operated induction hull valve. *(Plate 7, BR.1963/8)*

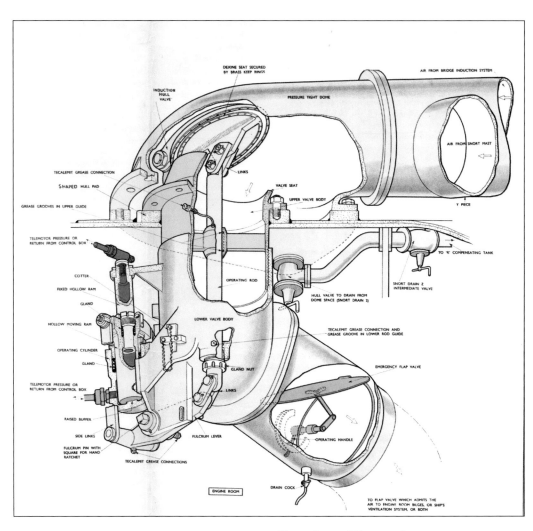

BELOW Starboard muffler valve grinding gear for removing carbon build-up on valve facings. *(Author)*

a telemotor-operated ram actuating a fulcrum lever and connecting rod located within the engine room.

Snort muffler valves

The function of the snort muffler valves was to shut off and isolate the snort exhaust mast when surface running. When running on the surface the exhaust gases from the diesel main engines are discharged through their respective group exhaust valves to the muffler valve located under the submarine casing and thence overboard out through the side of the casing. When snorting the diesel exhaust gases are diverted forward via the group exhaust valve to the snort muffler and thence to the standing exhaust mast. Operated manually from within the submarine, the muffler valve is provided with a hand-operated high-pressure air-blow facility to expel water from the snort exhaust mast prior to starting the diesel main engines, as water entering back into the diesel cylinders can have disastrous effects. The muffler valve was cooled by circulating seawater, the discharge of which 'dampens down' the exhaust, creating water vapour.

Diesel fuel oil and compensating systems

Diesel fuel oil is supplied from four external tanks (numbers 1 and 2 starboard tanks forward and numbers 3 and 4 starboard tanks aft) and from two internal tanks aft. The four external tanks have a combined capacity of 25,530 gallons (116,034 litres), the two internal tanks contain 16,697 gallons (75,997.9 litres), and there is a reserve tank of 14,370 gallons (65,311.7 litres), making the ship's total fuel oil capacity 56,597 gallons (257,233.4 litres).

From whichever tank, fuel passes to a four-valve cock-distribution chest where it is directed to either the diesel fuel oil centrifuge (see below) or a strainer. Fuel then passes to the 70-gallon (318-litre) gravity tank in the engine room which incorporates within it a snap tank from which it can then pass to either or both engines via a six-valve chest directing it to a strainer, a flowmeter, a booster pump supply and by-pass cock and booster pump to the fuel pump inlet cocks. Fuel under pressure is then discharged from the fuel pumps to a common rail via a block to the injector spray valves.

The snap tank, formed with conical ends and fitted with a gauge glass, provides instant measurement of the fuel consumption of each engine. This is ascertained by taking note of the time taken (in seconds) for either engine to consume a known quantity of fuel (one gallon), indicated by the sudden fall in fuel level in the gauge glass as the fuel passes the conical parts of the snap tank when in use.

As the consumption of fuel reduces the bodily weight of the submarine, to maintain equilibrium the volume of fuel is compensated by a continuous feed of seawater that is supplied from the engine circulating water system via a distribution valve chest, which directs water to the forward or after fuel tanks accordingly. Under limited pressure, the compensating water also provides positive suction to the initial fuel pumps.

Centrifugal separators (lubricating oil and fuel oil)

Two De Laval type 4014 centrifugal separators are fitted, one purifying diesel lubricating oil, the other purifying diesel fuel oil; both

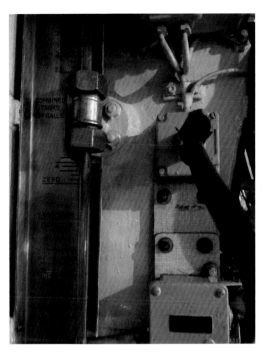

LEFT Diesel fuel oil snap tank with gauge glass indication to determine fuel consumption and efficiency. *(Author)*

machines are identical except for differences in their gravity discs.

If a liquid is allowed to settle within a tank, heavy particles in suspension will naturally gravitate to the bottom. If a liquid is subjected to a centrifugal force many times greater than gravity, separation is considerably accelerated and made more efficient. The degree of separation achieved is governed by the velocity at which the particles separate: the greater the speed of rotation of the centrifuge, the greater its efficiency in forcing the impurities radially outwards, at 90 degrees, towards the wall of the centrifuge cylinder.

The lubricating oil separator removes impurities – miniscule particles created by engine wear – as well as water. Removal of water is necessary as it reduces the efficiency of engine lubrication, but, more important, removal of seawater, which is used as a medium for cooling for the oil, purges the system of abrasive salt crystals that are detrimental to moving engine components. Contamination of the oil with seawater results from leaks, both within the lubricating oil cooler and within the engines from the water used to cool the cylinder heads and liners.

The fuel oil separator removes any compensating seawater used to pressurise the fuel tanks and to balance displacement of expended fuel.

In the two oil centrifugal separators, efficiency is maximised by use of a series of six conical disc plates mounted vertically upon the central driveshaft. These disc plates, made of steel, are manufactured to precise specification equal to the specific gravity of the fluid being purified in relation to the fluid or particles needing separation, the impure particles accumulating on the numerous surfaces of the discs. As impure fluids are discharged out of the machine, the decontaminated fluid that is left is passed back to its appropriate tank.

In both separators, for lubricating oil and fuel oil, the centrifuge is driven by a 1¼hp continuously rated electric motor through a safety friction plate and worm gearing. The friction plate ensures smooth running without overloading the motor. The worm gearing provides a speed ratio of 1:4 between motor and centrifuge bowl, producing an operating

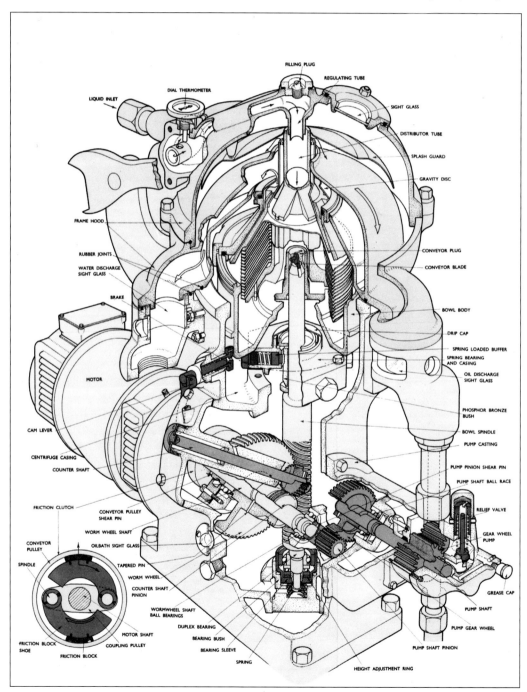

RIGHT De Laval centrifuge general arrangement.
(Plate 2, BR.1963/8)

speed of 6,000rpm and giving a running capacity of 250 gallons an hour (1,136.25 litres). In the lubricating oil separator, the oil is additionally passed through an electrical heater in order to maintain a temperature of 140°F (60°C) to optimise working viscosity, the temperature being monitored on the dial thermometer on the top of the separator.

Because the separators run at very high speed, integral sheer pins fitted in the friction clutch disengage drive if running speed exceeds 7,000rpm.

Maintenance of centrifugal separators

The following procedure for maintenance of centrifugal separators is taken from BR.1963 (8) Handbook for A Class Submarines Part 8 (Main Engine Associated Equipment) and personal practical experience.

Daily maintenance is essential to the smooth running and operation of the centrifugal separators. In particular, the separators do not function unless the water seal is maintained and this should be checked frequently. Any discharge of oil through the water discharge to bilge indicates failure and possible causes are:
- Insufficient sealing achieved on starting.
- Incorrect gravity disc fitted.
- Rubber sealing ring between bowl cover and bowl leaking.

Both separators need cleaning once every 24 hours or more often as required, such as after flushing through the main engines with lubricating oil after maintenance.

Cleaning procedure

1. Stop separator and allow speed to run down, then apply cam-levered brake shoe to stop centrifuge spinning.
2. When stopped, release centrifuge dome cover using wheel screw clamp. Lift back dome cover on its hinges to rest on its support bar.
3. Lock the purifier bowl body by equally turning both opposing radial locking pins, ensuring that they lock properly in the peripheral slots. Over-screwing with undue force will distort the spindle drive (hand-tight is sufficient).
4. Once locked, release the bowl body cover by unscrewing locking ring clockwise with a special ring key spanner (a little force may initially be required); remove locking ring.
5. Lift off conical top disc to release distributor tube and expose top of the intermediate disc plate.
6. Lift out distributor tube and carefully remove stack of 32 intermediate steel separator disc plates and place them in a bucket of diesel fuel for cleaning.
7. Clean distributor tube and the inside of the bowl body, taking care with the sludge space.
8. Clean all steel intermediate disc plates and inspect each for damage.
9. Carefully reassemble intermediate disc plates on to distributor tube, ensuring all are in correct numerical order (each are individually stamped) and aligned in their keyways.
10. Remount distributor tube into bowl body, ensuring the base locating dowel pin is fully engaged.
11. Replace conical top disc cover, checking cover is correctly located with the tongue piece in the bowl body.
12. Remove and inspect rubber sealing ring, checking it is free of damage (these can distort over time); reinsert or replace, and seat ring in its locating groove.
13. Replace locking ring and screw up hard using special ring key spanner.
14. Unscrew both radial locking pins, fully release shoe brake; check purifier bowl spins feely.
15. Close separator dome cover and secure with wheeled locking clamp.
16. Using special key tool, unscrew water filling plug and fully prime separator with fresh water, ensuring that excess water is seen passing out through water discharge. Check plug is firmly screwed up.
17. Start separator and check that diesel fuel oil/lubricating oil is seen flowing through the discharge sight glass; if not, reprime.

Battery and electrical distribution

Main battery

Manufactured by either Exide or DP Kathanode Tudor Battery Company, the main battery used in A-class submarines comprises 224 individual Exide lead-acid cells each weighing half a ton, providing an average pressure of 2 volts. Divided into two equal self-contained sections housing 112 cells connected in parallel, these provide a capacity of 6,630 ampere hours at a nominal output pressure of 220 volts DC. In construction each cell contains 30 negative plates and 29 positive plates interleaved with separators of ribbed wood or MIPOR (micro porous rubber). The electrolyte used is H_2SO_4 (sulphuric acid).

No 1 battery is located in No 1 battery tank situated below the central section of the accommodation space; No 2 battery (within battery tank No 2), is situated below the control room. Each battery tank is lined with Rosbonite to prevent potential acid corrosion to the submarine structure. Standing on waxed teak gratings the cells are arranged in two tiers and retained in place by rubber pads to prevent electrical contact between the cells and the submarine structure. The floor of each tank is inclined towards drain sumps. The clearance above No 1 battery is about 10in (25cm); No 2 battery about 3ft (0.9m). Access to each tank is gained through screw-down rubber-seated 'battery boards', which enable an electrical rating watch-keeper (refer to note) to take a 'battery dip' to determine the SG (specific gravity) of the H_2SO_4 electrolyte. Although naturally ventilated, each tank is fitted with exhaust fans, which are run when the battery emits gases during charging.

(Note: Electrical trained ratings, often referred to as 'greenies', were also colloquially nicknamed 'amp tramps'.)

It was essential that seawater ingress to the batteries never occurred. Not only would this cause serious damage to the cells, but the resultant chemical reaction with the electrolyte would also produce lethal chlorine gas similar to that used against ground troops during the First World War.

Battery charging

The following procedures and routines for battery charging are taken from BR.1963 (10) Handbook for A Class Submarines Part 10 (Electrics).

A. Standing charge
Prerequisites
1. No smoking throughout the boat (because of the presence of hydrogen).
2. One or both diesel main engine(s) driving one or both main motor(s) as motor generators.
3. One or both tail shaft(s) clutched in at sea and out in harbour.
4. Take readings of battery electrolyte density within specific gravity, temperature and voltage.
5. Sight battery sumps dry.
6. Start battery ventilation fans at full speed.
7. Check switchboard clear of earths.
8. Start main motor's cooling fans.
9. Check tail clutch OUT.

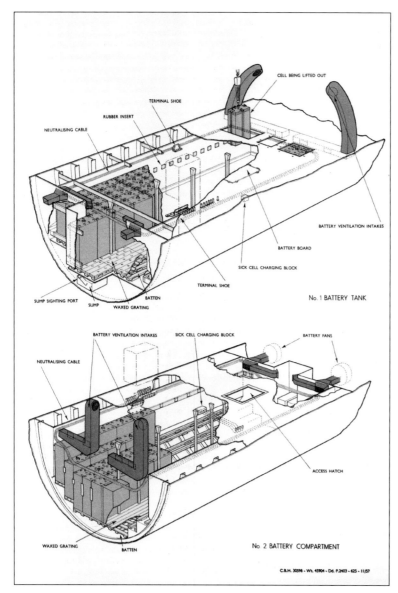

BELOW A-class battery arrangement (No 1 battery tank and No 2 battery compartment). *(Plate 2, BR.1963/10)*

10 Check engine clutch IN.
11 Check reversing switch to AHEAD and Grouper switch to UP.
12 Start the reducer putting reducer switch to REDUCING.
13 Put one switchboard voltmeter to supply; ie busbar voltage, the other to forward or after armature; eg generated voltage.

Procedure
1. Start main engine and adjust revolution to charging speed with engine running light.
2. Make the field switch and adjust strength so generated voltage exceeds supply voltage by 5 volts.
3. Make the starting switches.
4. Increase engine load keeping revolutions constant by increasing field strength.
5. Cell temperatures to be monitored throughout the charge.

Precautions
1. Charge to be immediately broken if snorting conditions dictate otherwise.
2. Charge to be immediately broken if battery parameters fall out of limitations.
3. Charge to be immediately broken if an earth warning develops.

B. Breaking battery charge
1. Stop main engine.
2. As charging current falls, and as engine stops, pull out the starting switches.
3. Adjust the field to fine field, break the field switch. Move field regulator to full field.
4. Put on tail shaft break.
5. Put in turning gear and line up tail clutch.
6. IN tail clutch, out engine clutch, out turning gear, main motor is now ready to propel.

C. Running charge
1. Carry out steps 1 to 5 and 9, 10, and 11 as for standing charge. The tail clutch and engine clutch will both be IN, then:
2. Make field switch and adjust generated voltage to 5 volts in excess of voltage supply.
3. Make the starting switches.
4. Increase engine load. The amount of power available for charging depends on power output of engine at revolutions ordered.

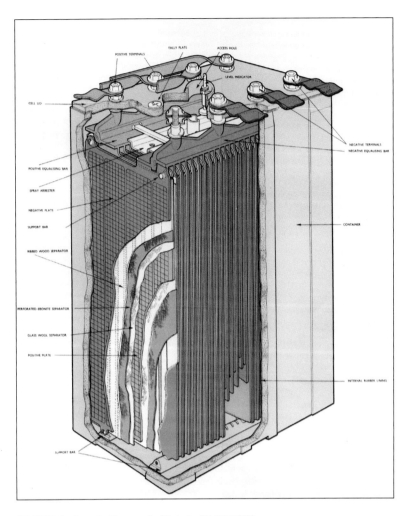

ABOVE A-class battery cell. *(Plate 1, BR.1963/10)*

D. Breaking battery charge on diving
1. Carry out steps 1 and 2 as for standing charge then:
2. OUT engine clutch.
3. Adjust field to full field.
4. Put the main motor to half ahead or grouped up and subsequently obey telegraphs as ordered.

Electrical distribution

While the main motors are supplied directly from the main switchboard, auxiliary equipment and lighting is supplied from two ring mains providing a voltage of about 190 volts when the battery is fully discharged, rising to 315 volts when fully charged. One ring main is of variable pressure, the other controlled pressure, the output supply of which is restricted by a machine called a reducer.

Chapter Five

Alliance's weapons

HMS *Alliance* and her type were among the most formidably armed submarines of the time and were well capable of meeting potential combat requirements. There were two weapon systems: torpedoes for submerged attack and guns for surface attack and defence.

OPPOSITE Embarking torpedoes or 'loading fish'. This is *Alliance*'s sister boat *Aurochs*.

ABOVE British Mk VIII 21in torpedo.

Torpedoes

For main armament the *Alliance* originally carried a total of 20 Mk VIII-type torpedoes, 10 of which were reloads stowed within the fore and after torpedo stowage compartments. In the late 1960s the Mk XXIV-type torpedo was developed and thereafter *Alliance* and other later classes of boat (the Ps and Os) would carry a mixed batch of torpedo armament depending on operational deployment.

Mk VIII torpedo

Differing little in design from the 18in torpedoes used during the First World War, the Mk VIII 21in torpedo was designed as the first British burner-cycle torpedo in 1925 and introduced into service in 1927. Powered by a self-contained diesel engine with integral air and fuel supplies, these weapons proved highly successful during the Second World War with some 3,732 having been fired by September 1944, this figure representing 56.4 per cent of all torpedoes fired by Royal Navy vessels during that period. Unsophisticated in terms of control devices, these torpedoes were simply pointed towards their target and fired. The specifications of the three developments of the Mk VIII torpedo are given in the accompanying table.

The Mk VIII torpedo was discontinued in 1983, after 56 years of service in the Royal Navy. As a brief digression, there were two most notable incidents where the Mk VIII was used with significant effect. First, the only wartime sinking of one submarine by another while both were submerged occurred on 9 February 1945 when Royal Navy submarine HMS *Venturer* (P68) sank the German U-boat U-864 with four Mk VIII torpedoes. Second, the only sinking of a surface ship by a nuclear-powered submarine in wartime occurred during the Falklands War when, on 2 May 1982, Royal Navy submarine HMS *Conqueror* (S48) sank the Argentine cruiser ARA *General Belgrano* with three Mk VIII torpedoes.

Mk XXIV torpedo

Named Tigerfish, this weapon is a heavyweight acoustic homing torpedo developed by Marconi during the late 1950s to provide a deep-diving torpedo with a speed of 55 knots

A-CLASS TORPEDO SPECIFICATION

	1927 Mk VIII	**'Early' Mk VIII**	**'Later' Mk VIII**	**'Early' Mk XXIV**	**'Later' Mk XXIV**
Weight	3,452lb (1,566kg)	3,452lb (1,566kg)	3,452lb (1,566kg)	3,414lb (1,550kg)	3,414lb (1,550kg)
Length	21.6ft (6.58m)	21.6ft (6.58m)	21.6ft (6.58m)	21.2ft (6.46m)	21.2ft (6.46m)
Diameter	21in (53.3cm)	21in (53.3cm)	21in (53.3cm)	21in (53.3cm)	21in (53.3cm)
Speed	41.0kts (47.2mph/72.0kph)	45.6kts (52.4mph/80.0kph)	41.0kts (47.2mph/72.0kph)	55.0kts (63.3mph/101.9kph)	24.0kts (27.6mph/44.5kph)[1]
Range	2.5 nautical miles (2.8 miles/4.6km)	2.5 nautical miles (2.8 miles/4.6km)	3.5 nautical miles (4.0 miles/6.5km)	7 nautical miles (8.1 miles/13.0km)	22 nautical miles (25.3 miles/40.7km)
Type of high-explosive charge	TNT	Torpex	Torpex	Torpex	Torpex
Weight of high-explosive charge	750lb (340kg)	722lb (327kg)	805lb (365kg)	295–750lb (134–340kg)	295–750lb (134–340kg)

[1] There was a brief capability to increase speed to 35kts (40.3mph/64.8kph), in which case the range was 7 nautical miles (8.1 miles/13.0km).

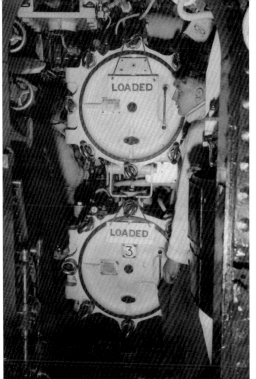

FAR LEFT HMS *Aurochs* after torpedo compartment with Mk XXIV torpedo on the rack to the right and Nos 5 and 6 torpedo stern tubes in the background.

LEFT *Alliance*'s Nos 1 and 3 starboard fore torpedo tubes with their 'loaded' tallies. Note how the ratings serve to illustrate the close confines of the working space.

(63mph/100kph). Propulsion was by an internal combustion engine, carrying high-pressure oxygen as an oxidant.

This weapon used a wire guidance system developed from that of the previous Mk XXIII torpedo. The wire guidance system controlled the torpedo by transmitting data from the submarine's sonar (in unison with the torpedo's autonomous active/passive sonar) along a wire connecting the launching submarine to the torpedo. This method permitted the torpedo to be launched as soon as a target was detected at long range, so that the torpedo could start to close the range while the precise course and speed needed to hit the target were calculated by the submarine's command centre and transmitted to the torpedo's sensors 'down the wire'. Furthermore, the torpedo could also be reassigned to another target or recalled. Typically, wire-guided torpedoes run at high speed to close the range (the approach speed) and then slow down to minimise self-generated noise interference with on-board sensors.

When initially introduced into the Royal Navy, the Tigerfish guidance system proved somewhat unreliable. The author's personal recollections of trials undertaken from HMS *Acheron* are that the wires tended to sever prematurely, giving rise to the need for torpedo-recovery vessels.

During the 1960s the wire guidance homing system of the Mk XXIV torpedo was simplified, so as to resemble that of the earlier Mk XXIII weapon. The propulsion unit was also improved, the internal combustion engine being superseded by an electric motor powered by a silver-zinc oxide battery. Unfortunately this reduced the speed of the weapon from 55 knots to 24 knots with a brief capability of 35 knots for the final-attack phase.

Torpedo tubes

When first built the A-class boats were fitted with a total of ten 21in torpedo tubes, six located forward (four internal, two external) and four located aft (two internal, two external).

LEFT Internal view of a torpedo tube. (Author)

109
ALLIANCE'S WEAPONS

RIGHT Loading a torpedo inside *Aurochs*.

BELOW After torpedo loading hatch with locking bar and loading rails. Note square sockets either side to receive hatch strong-backs fitted before diving to alleviate hull pressure strain. *(Author)*

The internal tubes were categorised as 'dry semi-slack-fit type' and each was made up in two sections termed as the inboard and outboard lengths. The inboard length was 110in long and manufactured of 0.375in steel plate rolled and welded along its seam, providing an internal diameter of 21.75in (55.3cm). The outboard length comprised two parts: a long section made of 0.5in steel plate and a short length of enlarged internal diameter – called the 'head space' – to which was fitted the hydraulically opened bow cap door.

The external tubes were categorised as 'dry close-fit type' and each consisted of three gunmetal sections bolted together with spigot joints. When initially built, the *Alliance* and her class had the outer bow doors opened by rod gearing operated from within the submarine, but this system was superseded by a high-powered ram on the front of each tube.

Embarking torpedoes ('loading fish')

Torpedoes were supplied from a submarine depot ship, ammunition ship or operating base. Whatever the source, very strict safety guidelines were observed when embarking weapons or explosives. These rules were implemented to avoid potential explosion, which could result in the loss of the submarine and of life, and to avoid damage that would render torpedoes useless.

Torpedoes were taken into the boat through either the fore or after torpedo-loading hatches, both of which are set inclined to the pressure hull to facilitate easy access. The first torpedoes embarked would be loaded directly into each of the torpedo tubes (with warheads pointing towards the rear end of the tube) and the remainder would be stowed securely on to the respective torpedo racks. It took between 20 and 30 minutes to load each torpedo.

The following procedure for embarking torpedoes is taken from SOPs (Submarine Operating Procedures) and author's observations.

Prerequisites

1 Crane with a competent operator.
2 Check all lifting strops with their 'belly bands', shackles, blocks and tackles (including lifting eye plates) are in sound condition and within test date.
3 Torpedo stowage compartment is clear of hammocks and unnecessary gear.
4 Confirm all main vents are locked shut with their harbour cotters fitted to prevent accidental opening.
5 Request duty officer – the Officer of the Day (OOD) – to run the low-pressure blower to blow round main ballast tanks to ensure that the boat is riding high on the waterline and that the lower lip of the torpedo-loading hatch is well above water. The OOD records boat draught mark readings complete with all tank contents in the submarine logbook, in order to monitor changes in buoyancy and weight.

RIGHT **Overhead torpedo loading rails, fore torpedo compartment.** *(Author)*

CENTRE **Reload torpedo on stowage rack in the fore torpedo compartment with torpedo loading hatch ladder in background.** *(Author)*

6 On completion, report to the OOD that boat is ready for torpedo loading hatch to be opened.
7 When all personnel have been informed that submarine integrity is about to be compromised, remove torpedo loading hatch strong-backs and open torpedo loading hatch.
8 Commence assembling torpedo loading rails: internally they are lowered down to waist height; externally they extend about 10ft along the casing.
9 Ensure torpedo stowage racks are rigged ready to receive torpedoes.
10 Lay out protective coir (coconut) matting at loading hatch rim and within the torpedo storage compartment where necessary.

Procedure
1 When the weapon comes over the loading rail, ensure that the two belly-band guide wheels engage in the rails.
2 Using steady ropes attached at each end of the torpedo, tip the nose of the weapon through the hatch.
3 When weapon passes down into the submarine, ensure it comes up against stops at end of rails.
4 The torpedo men then turn the weapon horizontal and take off the belly band, which is returned to the crane loader.
5 At same time the torpedo men attach strop of the wire hoist above the torpedo to lift the torpedo off the rails and then drop the front part of the rails down.
6 Lower weapon down to deck on to the skid bar to slide it over to either port or starboard side.

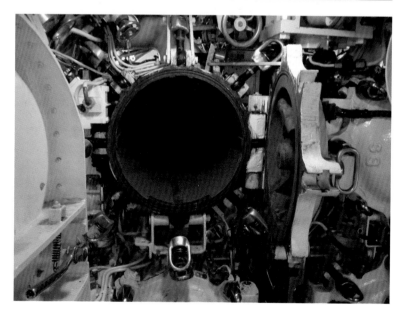

RIGHT **No 2 fore tube breech door in open position ready for loading with a 'fish' (torpedo).** *(Jonathan Falconer)*

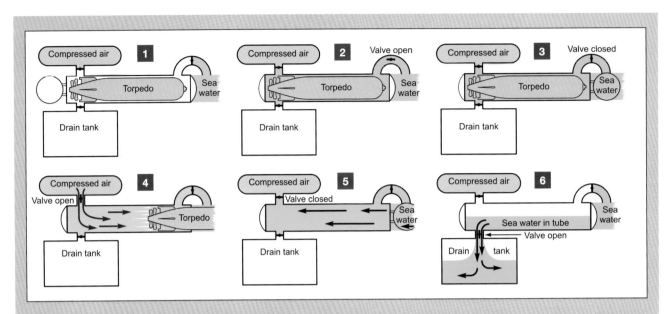

ABOVE Basic principles of firing torpedoes. (Roy Scorer)

Firing and reloading torpedoes

Before going on to the precise procedure for firing and reloading torpedoes, taken from SOPs, the basic principles can be summarised as follows:

RIGHT Fore ends: pressure gauge torpedo firing system HP pressure from main reducer. (Jonathan Falconer)

1. With the bow cap (muzzle) door shut, the torpedo tube breech door is opened and the torpedo loaded warhead first; the breech door is then shut and the bow cap door can then be opened safely.
2. Whatever the depth, sea pressure prevents the bow cap door from opening. To counteract this external pressure, an equal pressure is created within the torpedo tube by admitting water from the water-round torpedo tank. The displaced air is vented inboard and then a valve opened to the sea to equalise the pressure.
3. When the torpedo tube is flooded and pressure is equalised, the bow cap door can be opened and the torpedo is ready to fire.
4. Compressed air is then emitted into the torpedo tube, ejecting the torpedo from the tube. The air is not permitted to completely fill the tube, but vented off inboard so that a bubble does not escape into the sea and rise to the surface, giving away the position of the submarine.
5. The compressed air is then shut off and the torpedo tube fills with seawater, the weight of which offsets the lost weight of the torpedo to maintain the submarine in trim.
6. The bow cap door is then shut and a valve to the torpedo operating tank is opened to allow the tube to drain; the breech door can now be opened and the tube reloaded.

FAR LEFT Torpedo tube air charging main isolation valve. Note system pipework colour identification. *(Author)*

LEFT Fore ends: Nos 2, 4 and 1 and 3 bow cap operation levers. *(Jonathan Falconer)*

Procedure for firing a torpedo

Firing is done from the control room. With target identified, the captain will order, 'Action stations, action stations – torpedoes – attack team close up.' The procedure described below applies to Mk VIII and Mk XXIV torpedoes; the former travels in a direct line to its target while the latter is remotely guided from the control room.

Procedure

1. Torpedo tubes are flooded with seawater.
2. Internal pressure is equalised so that the torpedo tube bow cap doors can be opened.
3. At the same time torpedo tube air reservoirs (located underneath the torpedo tubes) are charged to 3,000psi from the high-pressure air ring main.
4. The UW2 (second-in-command in the torpedo compartment) mans the four firing levers and control box above the firing panel located in the centre of the torpedo compartment.
5. The UW2 watches the indicator lights from the control room – green means 'OK'.
6. Open outer torpedo tube doors by operating hydraulic delivery valve; check indicator reads 'open'.
7. When indicator is red, the UW2 pulls back the lever, allowing the air from the air reservoirs to go into the back of the torpedo tube and push out the torpedo.
8. As the torpedo exits the torpedo tube, the starter trigger lever fitted on top of the torpedo is tripped back, starting the torpedo's propulsion engine.
9. Torpedo warhead will only automatically arm itself after running 300yd (274m). This distance provides the submarine with a safety margin against premature explosion by weapon failure, the warhead detonating on direct impact with the target.

Procedure for reloading a torpedo

Reloading torpedo tubes with torpedoes weighing some 1½ tons apiece when at sea can be a precarious procedure involving considerable manhandling within a confined space and the difficulties can be exacerbated by motion in heavy seas. To this end reloading is best undertaken with the boat running submerged with activity noise kept to a minimum to avoid detection.

Prerequisites

1. Check torpedo tube bow (or stern) cap doors are shut and hydraulically isolated to prevent inadvertent opening, and lash operating levers.
2. Fore (or after) torpedo stowage compartment must be cleared of unwanted gear.
3. Shut and clip fore (or after) torpedo stowage compartment watertight bulkhead doors (a precautionary measure).
4. For fore torpedo tubes only, open the torpedo compartment port and starboard watertight doors, and tie and lash them clear.

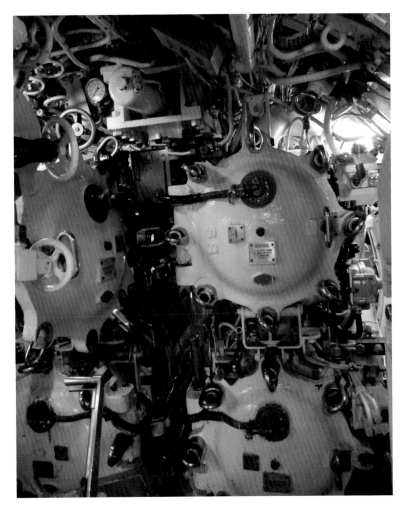

ABOVE Nos 2 and 4 torpedo (left) and 1 and 3 (right) fore torpedo tube breech doors shut and clipped. *(Author)*

BELOW Mk XXIII (QF) 4in deck gun from *Alliance*'s sister boat HMS *Andrew*. *(Author)*

5. Rig torpedo rails at waist height.
6. Rig lifting belly strops to reload torpedo together with lifting tackle and blocks.
7. Vent and drain torpedo tubes to be loaded and report to Officer of the Watch (OOW).

Procedure

1. Check torpedo tube to be opened is fully drained by means of the breech door test cock; when confirmed clear, inform control room OOW by reporting 'About open number X TT.'
2. Unclip the breech door by easing its Thetis clip and slowly open the door; report 'Number X TT open.'
3. Visually check bore of torpedo tube.
4. Using lifting strops and tackle, slowly manhandle and move reload torpedo towards torpedo tube breech, ensuring that the torpedo has been rotated with its trigger mechanism uppermost.
5. Slowly haul torpedo into the tube and when it is fully home shut and clip the breech door.
6. Repeat the procedure until all torpedo tubes are loaded.
7. When complete report, 'Numbers X, Y etc TTs loaded, all TT breech doors shut and clipped.'
8. De-rig compartment and rails, revert bow (stern) cap operating systems to normal, unclip and open torpedo stowage compartment watertight doors, and report.

Mk XXIII quick-firing 4in gun

Mounted at the fore end of the conning tower/bridge structure on an 'S2-type' mounting, this quick-firing weapon was primarily used for defensive purposes but could also be used offensively to target enemy shipping as an alternative to expending valuable torpedoes. In this case the submarine would surface to undertake the gun action, the boat being well trimmed down with her casing awash to minimise her waterline profile. The full specifications of the gun are given in the accompanying table.

Ammunition

Ammunition comprised high-explosive direct-action (HEDA) 4in (101.6mm) shells for flat

MK XXIII QUICK-FIRING 4IN GUN SPECIFICATIONS

General specifications	
Length	137.6in (349.4cm)
Weight with BM, unloaded	1,569lb (711.7kg)
Weight without BM, unloaded	1,457lb (660.9kg)
Calibre (bore diameter)	4in (101.6mm)
Length of bore	132.16in (335.7cm) or 33.04 'calibres'[1]
Length of rifling	116.58in (296.1cm)
Rifling (1 turn in 25 calibres[1])	32 grooves
Breech	Horizontal sliding block
Muzzle velocity	1,750ft per second (396m per second)
Maximum range	9,700yd (8,870m)
Gun	1,569lb (711.7kg)
Gun cradle and fittings	1,492lb (676.8kg)
Gun carriage and fittings	20lb (9.7kg)
Base	1,426lb (646.8kg)
Gun shield and platform[2]	1,279lb (580.1kg)
Sighting gear	367lb (166.5kg)
S2 mounting, other specifications	
Limits of elevation	+30° to −10°
Training gear ratio	One turn of handles to 2.95° of training
Elevating gear ratio	One turn of handles to 3° of elevation
Initial compression on run-out springs	1,585lb (718.9kg)
Final load on run-out springs	3,170lb (1,437.9kg)
Recoil, working length	24in (61.0cm)
Recoil, metal to metal	26in (66.0cm)
Capacity of recoil buffer and make-up tank	14.5 pints
Liquid in system (oil)	DTD 44D (OM1)
Force incurred on recoil	14.2 tons
Upward lift of mounting	10.0 tons
Downward blow	12.25 tons
Diameter of base of mounting	60in (152.4cm)
Holding-down bolts	24 of 1in (2.54cm) diameter
Bolting arrangement	Pitch circle diameter of 57.5in
Sight data, range gear	
Maximum range angle	20°
Dial	Calibrated to 9,600yd
Gearing ratio	16.571428:1
Angle of drift	20°
Sight data, deflection gear	
Maximum deflection	55 knots at 2,000yd
Maximum angle of deflection	3° 3'
Gearing ratio	5.55:1

[1] Calibres = length of bore x diameter of bore
[2] Includes ammunition trough and range/deflection receiver supports and brackets

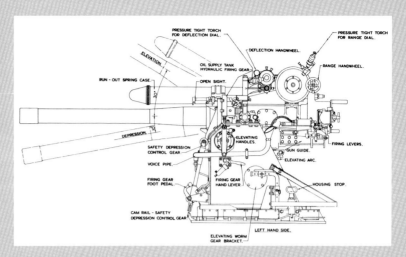

ABOVE Mk XXIII (QF) 4in deck gun left-hand side view.

BELOW Mk XXIII (QF) 4in deck gun right-hand side and rear view drawings. *(Both BR.1845/March 1969)*

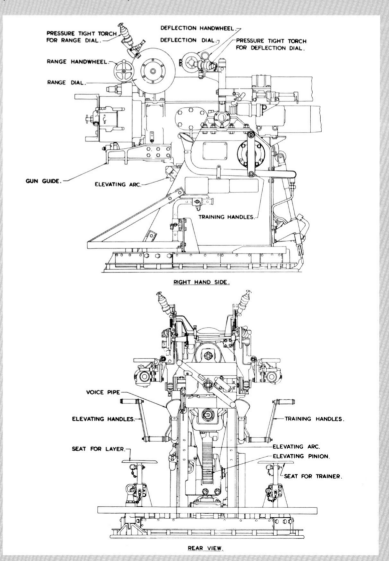

trajectories, designed to burst with great shattering effect on unarmoured targets but to have little or no penetration. These shells contained the largest possible bursting charge consistent with the shell having the necessary strength to withstand the shock of discharge from the gun. Fitted with nose fuses, they were mainly used against ship targets or for bombardment. The projectile weighed 35lb 13oz (16.1kg) and the weight of the TNT explosive charge was 3lb 12oz (1.9kg): TNT is less hygroscopic (water absorbent) than the explosive Lydite and therefore preferable for use at sea.

Operation and maintenance of Mk XXIII quick-firing 4in gun

The following procedures for gun operation and maintenance are taken from: BR.205/47 Drill for 4-inch Mark XII, XII* and XXII on SI Mountings (Submarines), 1947 (HMS *Excellent* Museum); BR.968 1925, Handbook for 4-inch QF Mark XII gun on SI Mounting (HMS *Excellent* Museum); BR.1845, Handbook for 4-inch QF Mark 23 gun on 4-inch S2 Mounting (1949, Priddy's Hard collections, Hampshire County Record Office, Winchester).

The gun is to be fully overhauled every three months by an ordnance artificer (OA).

RIGHT *Alliance* crew members in front of the QF 4in deck gun, HMS *Alliance*.

Before proceeding to sea
1 All fittings including sight bracket and breech mechanism are to be in place.
2 Telescopes only are to be removed.
3 Non-working steel components – ie gun pedestal arms of carriage, side bar, limiting pawl, steel plate of cradle box and controlling unit – are to be coated with a compound mixture comprising one part of white lead and two and a half parts of tallow.
4 Breech mechanism, breech block, lock, levers of firing mechanisms, and sights to be lightly oiled.
5 Bore of gun to be lightly oiled (note that heavy oiling may cause the shell to break up in the gun); Vaseline or heavy oil is not to be used.

When at sea
This is very much dictated by operational restrictions or opportunities of surface running in daytime.
1 Work gun in elevation and training once every 24 hours.
2 After prolonged diving strip breech mechanism and work firing gear.
3 Remove all traces of rust from non-working steel parts and treat with aforesaid compound of white lead and tallow.

Gun drill for Mk XXIII quick-firing 4in gun

The following procedures for gun operation and maintenance are taken from the same sources as 'Operation and maintenance of Mk XXIII quick-firing 4in gun'.
The gun crew comprises seven men whose titles, duties and manning positions are as follows:
- **Gunlayer (1)** Gun captain, wholly responsible for the gun; stationed at the elevating hand wheel.
- **Breechworker (1)** In charge of loading the gun and safety precautions at the breech; stands to the right-hand side in line with the breech and facing it.
- **Loader (1)** Stands to the left of the breech, clear of recoil and facing the breechworker.
- **Trainer (1)** Stands at the training hand wheel.
- **Sightsetter (1)** Stands at the sights.
- **Ammunition supply (2)** Two men who

pass ammunition and stand at the top of the gun tower; they are supported by further personnel (usually seamen and stokers) forming an ammunition train from the magazine within the submarine up through the gun tower.

Stand by for gun action

The following procedures for gun action are taken from BR.205/47. To bring the gun into action immediately on surfacing, when the enemy has been sighted while the submarine is submerged, the gun's crew muster at the foot of the gun tower and with the necessary equipment. The gunlayer reports, 'Gun's crew present and correct.'

The following orders are passed while the boat is still submerged:
1 'Bearing and description.'
2 'Range and deflection' (sightsetter writes this on his wrist slate).
3 'Point of aim.'
4 'Salvos shoot.'
5 'Man the gun tower.' The gun's crew man the tower in the following order: gunlayer, with telescope (port side); trainer, with telescope (starboard side); loader, with firing lanyard and loading glove (port side); breechworker, with spare lock in canvas bag (starboard side); and sightsetter. As soon as the crew are in the tower, the ammunition-supply men pass up two rounds that are placed in ready-use clip lockers by the gunlayer and the trainer.
6 'Stand by to surface' (gunlayer and trainer remove the pins). On seeing gun tower manned, gunlayer reports, 'Tower manned, pins out.'
7 'Surface', followed by 'Off one clip.'
8 Gunlayer and trainer remove one clip and open the hatch. As the boat breaks surface, the gun's crew ascend to the casing and the gun.

Procedure for gun action

1 Signal for gun action given by the blowing of a whistle by the gunnery control officer (GCO) on the conning tower bridge.
2 Gunlayer removes the training stop, clears the sights bore and 'ships' (fits) his telescope.
3 If the sightsetter is not yet in position, gunlayer sets the sights.
4 Trainer unships the lock and places it in the spare lock stowage.
5 Trainer trains the gun in by open sight, ships his telescope and then trains accurately.
6 Breechworker takes a round from the loader, loads the gun and closes the breech.
7 Breechworker takes the lock from the canvas bag, ships it, fits the latch to 'Fire' and reports 'Ready.'
8 Breechworker takes a round from the loader and stands by to reload. The gun is now ready to fire. The firing order given by the GCO is 'Shoot.'
9 As soon as the round has been fired the breechworker repeats the procedure, with sighting and range adjustments made accordingly throughout. It should be noted that the casing around the gun was likely to be awash during firing and seawater probably would have been entering the gun tower.

Procedure to stop loading

1 'Stop loading' order given.
2 Gunlayer clears the gun (if loaded).
3 Gunlayer and trainer continue to follow the target.
4 Sightsetter continues to set the sights.
5 Breechworker and loader stand by to load.
6 At the order 'Salvos', gun is to be loaded as if empty; breechworker puts latch to 'fire'.

ABOVE The 4in deck gun in action on the S-class submarine HMS *Sibyl*.

Discontinuing action

The GCO orders 'Cease fire.' The gun is then cleared away as follows:

1. Breechworker puts latch to 'safe' and reports 'Gun loaded half cock' or 'Gun empty.'
2. If the gun is empty, it is left empty.
3. If the gun is loaded it is to be treated as a 'misfire' until the breechworker has put the latch to 'safe', seen that the striker has not moved forward, and reported a 'half cock'.
4. Gunlayer orders 'Out cartridge.' Breechworker opens breech and removes cartridge.
5. Loader and ammunition-supply personnel return ammunition to ready-use clip lockers (or magazine below).
6. Gunlayer and trainer train gun to securing position (ie, barrel muzzle facing forward).
7. Gunlayer replaces training stop.
8. Gunlayer and trainer unship telescopes.
9. All crew get below, taking all equipment with them.
10. Gunlayer is responsible for gun tower hatches being shut and clipped, and reported as such. The gun is now in the 'cleared-away position'.

RIGHT Sighting device on the Oerlikon 20mm AA gun. *(Author)*

Oerlikon 20mm anti-aircraft gun

When running on the surface submarines are most vulnerable to air attack by bombing, machine-gun fire or cannon fire from fighter aircraft. Defence against this is provided by a 20mm Oerlikon automatic gun mounted upon the 'bandstand' at the after end of the conning tower/bridge structure, a position that

OERLIKON 20MM ANTI-AIRCRAFT GUN SPECIFICATIONS

General specifications	
Weight with splined barrel	141lb (64.0kg)
Weight without splined barrel	150lb (68.4kg)
Weight of shoulder piece with 300kt sight	32lb (14.5kg)
Length overall	96.0in (243.8cm)
Calibre	20mm (0.787in)
Rifling	Nine right-hand grooves
Muzzle velocity	2,725ft per second (830m per second)
Rate of fire (when automatic)	465–480 rounds per minute
Maximum range at 45°	6,250yd (5,712m)
Effective range	1,000–1,200yd (915–1,100m)
Mk 12a mounting specifications	
Maximum elevation	80°
Maximum depression	12.5°
Muzzle sweep (radius)	50.0in (127.0cm)
Height of trunnions above deck	55.5in (139.7cm)
Horizontal distance between gun centres	7.5in (19.1cm)
Weight of complete mounting (minus ammunition)	2,334lb (1,059kg)
Outside diameter of base	32in (81.2cm)
Holding-down bolts	10 of 0.75in (1.9cm) diameter
Holding-down arrangements	Pitch circle diameter of 2.25in (5.7cm)
Ammunition	
Magazine capacity	60 rounds
Weight full	61lb (27.7kg)
Weight empty	30lb (13.6kg)
Nominal weight of one round	80.125oz (2,271g)

provides a wide arc of fire across the stern of the boat. The full specifications of the Oerlikon are given in the accompanying table, and as this is a gas-operated automatic weapon no formal gun drill is described.

Additional armament and ammunition

Guns

The *Alliance* was equipped with a Vickers gas-operated .303 machine gun that could be mounted on the port side of the conning tower.

While stationed in Singapore during the Indonesian/Malay uprisings in 1962–63, *Alliance* was also armed with a small-calibre deck gun for the duration of hostilities and covert operations.

When the A-class boats were initially under construction during April 1945, design proposals were prepared to mount a second gun between the after side of the conning tower and the engine room hatch. This intended feature, together with the existing number of torpedo tubes, would have made the A-class boats the most heavily armed submarine type in the Royal Navy since the advent of the K-class boats. However, as the war was clearly drawing towards its conclusion at that time, the concept of fitting an extra gun was abandoned on the grounds that it was both unnecessary and expensive.

Mines

During the Second World War submarines laid a considerable number of the Mk XVI-type mines. The type of mine that could be laid by A-class submarines was the magnetic Mark V acoustic type, which weighed two tons and was charged with 2,000lb (900kg) of explosive.

Mines were embarked into the submarine in exactly the same way as torpedoes and loaded into the tubes. When the submarine was operating as a minelayer, the mines were discharged from the torpedo tubes by high-pressure air in exactly the same manner as firing a torpedo.

Ammunition

Ammunition for the 4in gun, the Oerlikon and other weapons was stowed in the magazine located under the wardroom adjacent to the auxiliary machine space. Post-1955 the

ABOVE The range hand wheel (black wheel, right) and range dial (steel and bronze wheel, centre) on the QF 4in deck gun. *(Author)*

magazine was reduced in size and converted into a sound room to accommodate the 187 sonar system. A list contained within the Ship's Book c.1947 records the following types and quantities of ammunition carried before removal of the 4in gun and Oerlikon:

- High-explosive direct-action (HEDA) 4in shells – 230 rounds.
- Vickers .303 – 4,875 rounds.
- Tracer – 3,750 rounds.
- Incendiary (B7) – 1,875 rounds.
- .303 rifles – 300 rounds.
- .38 pistol – 150 rounds.
- Type 36 grenades – 24 in number.
- Line towing rifle – 96 rounds.
- 2½in pistol grenades – 18 in number.
- 1in aiming rifle – 192 rounds.
- Flares – 10 of each colour (red, yellow, green).

The following list dates from 9 May 1970 and is the 'Initial Outfit of Ammunition, Armaments and Pyrotechnics for Commission':

- Sub-machine gun (9mm) – 2,500 rounds.
- Line-throwing rifle – 120 cartridges.
- Sub-machine gun magazines – 36 in number.
- 1½in signal cartridges – 18 in number.
- Flares – 10 of each colour (red, yellow, green).
- 36mm grenades – 24 in number.
- Stirling sub-machine gun – six in number.
- Line-throwing rifle – one in number.
- 1½in Very pistol – two in number.

Chapter Six

Alliance's operational equipment

When submerged, HMS *Alliance* was controlled by a conventional rudder and by hydroplanes fore and aft, and her crew observed activity on the surface by means of two periscopes with 'search' and 'attack' functions.

OPPOSITE Starboard forward hydroplane. *(Author)*

ABOVE **Starboard view of the after end of *Alliance* showing rudder and hydroplanes.**

BELOW **Steering gear schematic.** *(Plate 1, BR.1963/6)*

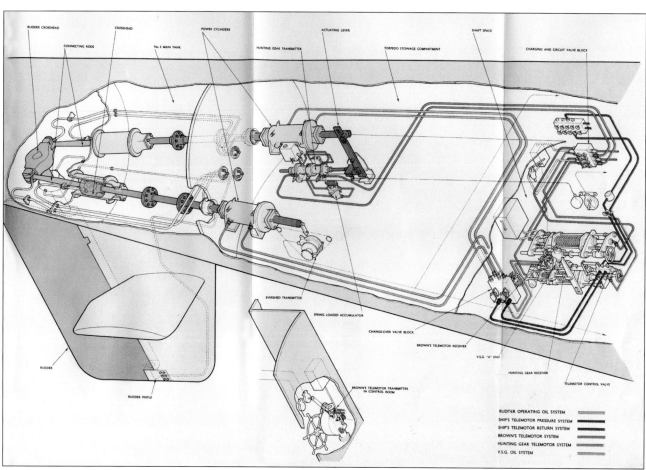

Unlike ships, submarines are controlled in two planes by virtue of operating while submerged.

Steering a conventional vertical rudder rotates the boat about its longitudinal axis in order to turn or manoeuvre to either port (left) or starboard (right).

Planing horizontally fixed hydroplanes, which pivot about their axes, determine the pitch of rise or dive of the bow and stern in the process of submerging or surfacing the boat. Once the boat has submerged and attained neutral buoyancy, the hydroplanes are used to control depth and trim (ie manoeuvring the boat about its horizontal axis). Functioning like horizontal rudders, hydroplane surfaces act upon the seawater in a similar way as the ailerons and flaps of fixed-wing aircraft.

Steering gear and control systems

The steering is operated by hydraulics, the power source being a motor-driven pump that runs at a constant speed but gives variable delivery by means of variable speed gear (VSG), with output governed in both direction and quantity from the Brown's telemotor system in the control room.

The rudder is moved by two hydraulically operated pistons housed in power cylinders that are connected to the rudder crosshead by actuating and connecting rods, between each of which is installed a crosshead that provides the necessary articulation movement needed when the rudder is in any position other than centralised. As it is necessary to have a clear working area for the after torpedoes, the VSG and Brown's telemotor system receivers and transmitters are located in the shaft space. This arrangement requires a telemotor hunting gear system to be incorporated between the power cylinders and VSG control gear.

Because steering is an essential operating system, it is backed up with an alternative oil supply from the submarine's telemotor system via a changeover valve block that effectively bypasses the VSG. Direct operation of the rudder by VSG or indirectly from the submarine telemotor system is governed from the control room or locally within the shaft space. Should local control be necessary, a gyro repeater is fitted within the shaft space. Should both VSG and submarine telemotor systems fail, a hand pump is fitted in the after ends of the boat to provide power to the operating rams.

Fitted directly behind the downward-projecting tail fin, the fabricated-steel rudder weighs about 4.17 tons and has an effective working area of 60sq ft. The entire component is free-flooding to avoid buckling under pressure differentials when wholly submerged.

For safety reasons *Alliance* can be steered by six modes, the first three a matter of routine, the remaining three when defects arise.
- Normal.
- Emergency.
- Silent routine.
- Steering by variable speed gear (VSG) in local control.
- Steering by telemotor system in local control using the hunting gear.
- Steering by telemotor system and hand pump.

LEFT **Control room showing helm (steering) control station with steering lever, rudder position indicator and gyro repeater.** *(Jonathan Falconer)*

Normal steering

In normal mode the steering is controlled from the control room using the Brown's telemotor system with the VSG running continuously. The VSG supplies hydraulic power to the rudder actuating system with the system changeover valve block set to connect the VSG to the power cylinders and shut off the submarine telemotor system.

Emergency steering

The objective with emergency steering is to initiate an immediate changeover to a different system while the failure or defect is investigated. The VSG is stopped and the system changeover valves are set to isolate the VSG and connect the submarine telemotor system on line to the power cylinders, with rudder movement being effected by means of a control hand wheel.

Silent routine steering

This is used to eliminate noise from the pump, steering being controlled from the control room. Here the changeover valves are set to cut out the pump and connect the telemotor control valve with the power cylinders. The submarine telemotor system return and pressure valves to the control valve are opened. In this mode the Brown's telemotor system operates for normal steering, but moves the submarine telemotor control valve, and the hunting gear centralises the rudder.

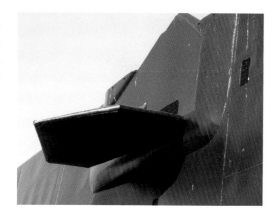

RIGHT Starboard fore hydroplane set down into operating position. *(Author)*

Hydroplane gear

Acting like horizontal rudders, the hydroplanes control the submarine in the vertical plane, enabling the boat to rise or dive. Two sets of hydroplanes are fitted, forward and aft: the fore planes control depth and the after planes control the horizontal angle of the boat. Both sets of planes are hydraulically operated from the control room direct from the submarine telemotor system to the actuator rams and respective crossheads. In the event of failure they can be operated locally by a hand gear.

The fore hydroplanes are mounted above the waterline and fold into the vertical position to avoid damage when in harbour or in close proximity to other vessels; this facility also protects the tilting gear when in a heavy seaway.

The after hydroplanes are always submerged and fixed, and their position makes them well placed to take advantage of propeller wash. Being constantly submerged, they are highly vulnerable to damage and are therefore protected by a D-shaped steel guard. Two types of construction are used in the A-class: cast steel with a skin of 8lb plating; and a fabricated steel frame with a skin of 17lb plating.

During manufacture cast-iron weights were fitted to balance the hydroplanes about their shafts, and all internal surfaces were protected with a coating of a bituminous solution.

Because the after planes control angle, it is essential that they can be brought into a neutral position to safeguard the submarine. To facilitate this, three eye plates are welded to each plane and 2in steel wire ropes can be secured to these eye plates with shackles. When required, this work had to be carried out by a diver who

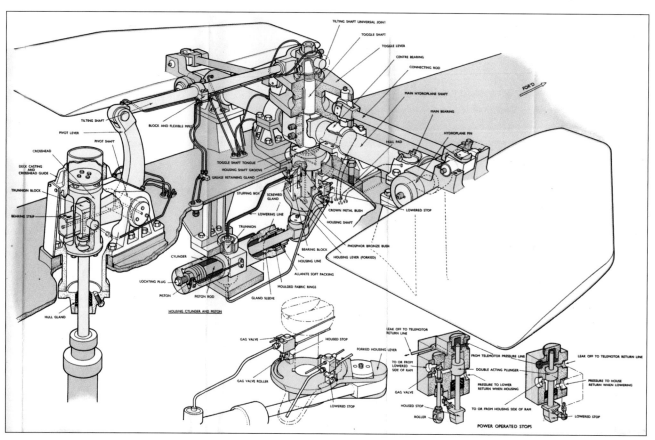

BELOW General arrangement of forward hydroplane housing and tilting gear. *(Plate 1, BR.1967/7)*

ABOVE General arrangement of after hydroplanes. *(Plate 4, BR.1963/7)*

BELOW Skeg (or rudder fin) supporting the starboard after hydroplane and rudder. Note the horizontal plane guard projecting from the hull. *(Author)*

would have led the wires over the after end of the plane guard through a bull ring; the wires would then have been hauled upon until the planes were set in a neutral position, the wires being secured to the nearest bollards.

There is one further point to mention about use of the after planes. When a submarine is turned at speed while submerged, its rudder will tend to drag down the stern and therefore immediate countermeasure action must be taken by the after planes operator to mitigate this effect.

Periscopes

The submarine's periscopes provide visual surveillance above the surface when the submarine is dived. *Alliance* has two periscopes, both of which pass from the control room through the pressure hull and up through the conning tower/bridge structure.

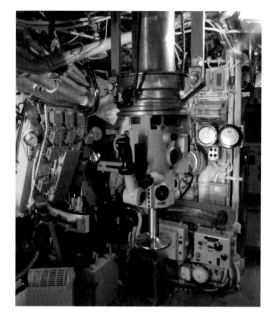

RIGHT Type CH66 attack periscope. *(Author)*

FAR RIGHT Type CK14 search periscope (note the hoist wire). *(Author)*

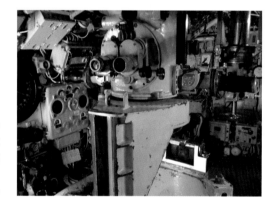

RIGHT Search periscope with attack periscope forward (right). *(Jonathan Falconer)*

The after 'search' periscope (type CK14), used for general vision, is binocular and the larger of the two. The foremost periscope (type CH66), used for attack purposes, is monocular and, by virtue of its function, obviously smaller and less conspicuous when protruding above the surface of the sea.

Periscope specifications are given in the accompanying table. Periscope functions are five in number as follows:

- **Sky search** For detecting airborne danger, by tilting the top prism.
- **Magnification** In high power this is 6X but in low power a minifying telescope is swung into field and reduces magnification to 1.5X.
- **Range estimator** Both periscopes have a range estimator that measures the angle subtended by a known height and introduces a ghost image (ie a second image of the target) superimposed over the actual image (further explained below).
- **Bearing reading** A bearing scale is fitted under the hull gland and a vertical line on the main periscope main tube indicates the bearing on which the periscope is trained. This scale is aligned with the torpedo tubes when the submarine is docked at each refit. When taking a bearing, the periscope is aligned on a target by a graticule etched on the lens in the top of the optical system.
- **Eyepieces** Focusing to suit individual observers' requirements is provided at the eyepieces and coloured screens of red, grey or polarised glass can be introduced into the line of vision. In the binocular periscope the interocular distance can also be adjusted.

PERISCOPE SPECIFICATIONS

	After search periscope (CK14)	Fore attack periscope (CH66)
Diameter	9.5in (24.1cm)	7.5in (19.1cm)
Vision	Binocular	Monocular
Magnification, high power	6X	6X
Magnification, low power	1.5X	1.5X
Field of view, high power	6°	10°
Field of view, low power	24°	40°
Graticules[1], distance apart, high power	0.25°	0.25°
Graticules[1], distance apart, low power	1°	1°
Sky search range	54° elevation to 15° depression	30° elevation to 15° depression
Height between eyepiece and top prism	40ft 3in (12.27m)	40ft 0in (12.19m)

[1] Graticules are divisions marked on the lens in degrees.

Periscope tubes and bearings

A high-tensile bronze tube provides the main strength for the periscope and is supported at its bottom end by a crosshead. The tube passes through the pressure hull by means of a Blockhouse gland and bearing that are supported externally by two bearings located in the periscope standard. The upper bearing in the periscope standard comprises a ball race. The lower bearing can be adjusted by wedges to line up with the other two bearings. The periscope is raised and lowered by means of wires operated by a telemotor press.

The main tube assembly of both the CK14 and CH66 periscopes is made in five lengths as follows:

- **Head** Containing the sky search and power-change mechanisms.
- **Top forging** Houses the first objective lens (CK14) or the first and second objective

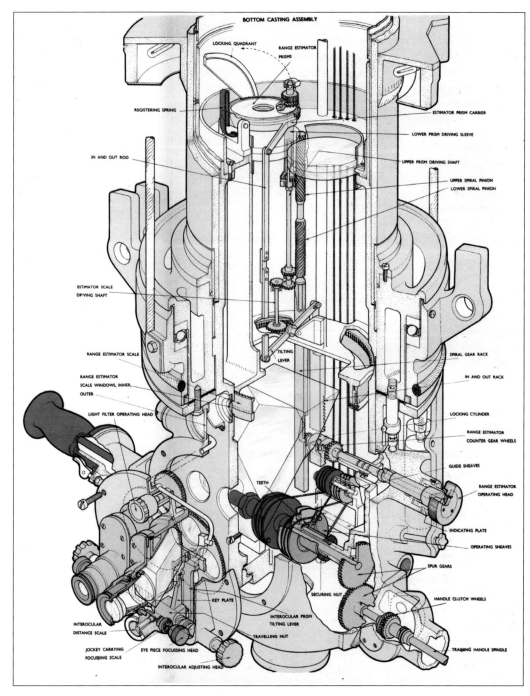

LEFT Type CK14 main tube assembly.
(Plate 1, BR.1963/11)

RIGHT Type CH66 main tube assembly.
(Plate 3, BR.1963/11)

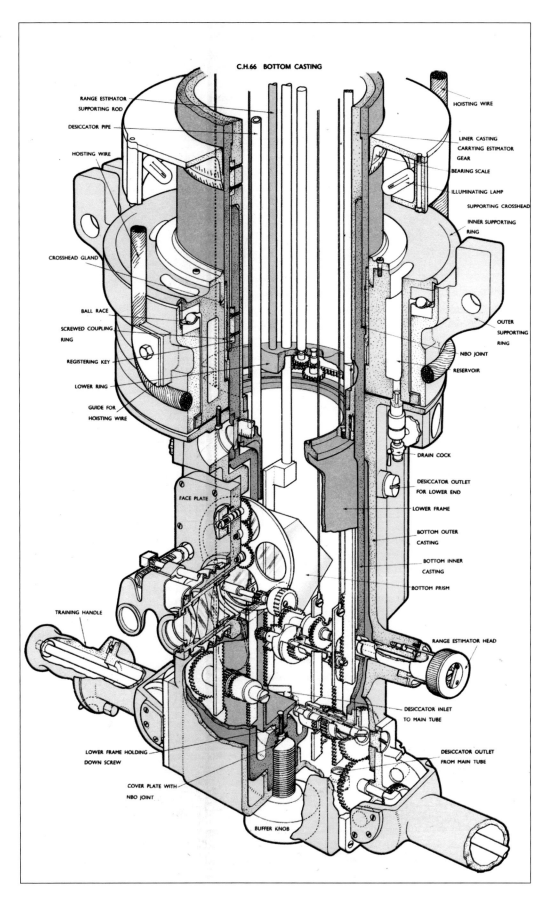

lenses (CH66) plus the combined graticule and deviating prisms.
- **Upper main tube** Carries the upper main objective lens.
- **Lower main tube** Carries the lower objective lens.
- **Bottom casting** With its integral crosshead, this contains the bottom prisms, the eye lenses, and all controls for magnification, power-change and sky search controls. The actual crosshead is suspended by wires, which take the entire weight off the main tube.

Range finder

The range finder fitted to the CH66 attack periscope comprises a pair of annular-shaped, wedge-section prisms. These can be brought into the path of light between the main lower objective lens and the bottom prism. While a normal image is seen through their central hole, a ghost image, displaced vertically from the normal, is created by the prisms. Displacement of the ghost image depends on the relative angled positions of the two prisms.

Range estimator

Applicable to both periscopes, the range estimator comprises a pair of annular-shaped, wedge-section prisms inserted into the right-hand field between the lower main objective lens and the bottom prism. Here the normal image is seen through the hole in their centre and a ghost image vertically displaced from the normal image produced by the prisms. The displaced ghost image depends fully on the relative angular position of the two prisms. By turning the prisms in opposing directions, the base of the ghost image can be made into a superimposed second image of the target over the actual image. To create this the captain adjusts the split prism so that the waterline of the second image is set on the masthead of the actual target image. Rotation of the prism can be read off a scale (graduated in minutes) giving the subtended angle by the height of the target at the eye of the observer. The height of the masthead from the water is entered on the dial, and the reading obtained. What are actually measured are angles, not distance. If the masthead height is entered accurately, the range will be correct. Getting the masthead height wrong gives an incorrect range.

How the optical system works

The description in this section applies to the CH66 monocular attack periscope but the principles are the same for the CK14 binocular search periscope.

Light from the target passes through the top window and is reflected down the tube by the top prism, which itself can be rotated about its axis to elevate or depress the line of sight from the horizontal. Light from the top prism

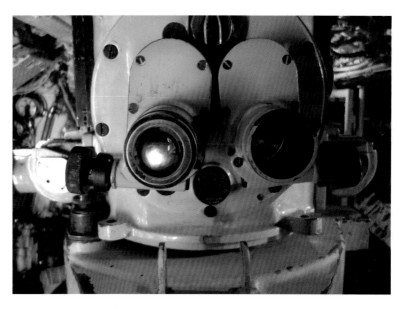

ABOVE Control room: eyepieces of the search periscope. Note the adjusting and control knobs on either side. *(Jonathan Falconer)*

BELOW Transatlantic passenger liner SS *America* seen through *Alliance*'s periscope.

then passes through a reversed Galilean low-power telescope comprising two lenses, the upper negative, the lower positive. The resulting effect of these lenses diminishes the size of the image, providing a total magnification of 1.5X at the eye lens. If the option of magnification of 6X is required, these two lenses are swung out of the path of light and therefore the light passes to the high-power first objective lens. This lens then forms an image of the object at the graticules (graduated divisions), and subsequently the etching on the graticule appears superimposed on the image seen through the eye lens. When viewed in high power this etching is a vertical line divided into ¼-degree steps by short horizontal lines with the intermediate-degree positions dotted in; when in low power the values become 1-degree and ½-degree markings respectively.

Because of its long, narrow top section, this type of attack periscope consists of three telescopes placed end to end, with the second image being formed between the third objective lens and the upper objective lenses. This image, which passes between the upper and lower objective lenses as a parallel beam, converges through the combined bottom prism and the collector lens that forms the third image. The observer then sees this image through the adjustable eye lens.

To suit individual interocular distance, the distance that the light beams emerge from each eye lens must be adjusted, by swinging the rhomboid prisms together or apart.

Mounted between the collector lens and the eye lens is a colour disc that can be rotated, a process that brings four filters – clear, red, grey and polarised – into field in sequence.

Anti-reflection film is deposited on optical surfaces to reduce the amount of light reflected, thereby increasing the amount of light transmitted. The only exceptions to this are the outer surfaces of the top window and the eye lens.

Any moisture present within the periscope assembly can obscure or distort the field of visibility. To prevent this, the main tube assembly of the periscope is furnished with a desiccation system comprising a series of fine copper pipes through which dry air is circulated; the main inlet and outlet connections are located at the back of the bottom casting. A separate desiccation system is provided for the face-plate assembly on the bottom casting.

Periscope maintenance

In harbour the periscope can be stowed in three modes:

- Fully down. In this position a wooden cap is fitted to protect the top bearing, which is not to be left exposed under any circumstances; failure to do this will necessitate a complete bearing change, which will require the removal of the entire periscope.
- Raised and on locked pins to effect cleaning of the damp well.
- With the bottom just clear of the periscope well, as when charging the desiccation system; this is the best position for the periscope to withstand any vibration.

Cleaning the top window

The top window at the head must be kept clean and when in harbour a linen protective hood is to be fitted over it. When cleaning to remove all grease and salt no abrasive materials are to be used. To avoid risk of scratching, lint-free cloth dampened with a suitable spirit solution is to be used. Traditionally the 'outside wrecker' – the nickname given to the artificer who maintains all auxiliary machinery, mechanical services and systems not associated with the engine room propulsion machinery – obtains gin from the wardroom for this purpose, gin being the purest spirit kept on the boat. If salt crud proves stubborn, a very soft wood scraper can be used. These rules also apply to the lenses of the eyepiece, etc, at the lower end of the periscope.

Maintaining the main tube

When dry after a period in harbour, the tube is to be wiped clean with fuel oil, lightly oiled and inspected for damage. The fact that the periscope tube is kept completely round cannot be over-emphasised, especially as the portion above the hull gland is highly vulnerable.

Periscope wells

These are to be well cleaned and dried regularly. When a man enters into the well for this task, the periscope is to be 'locked on its pins' and hydraulics isolated to avoid inadvertent periscope movement and injury.

Hoist wire adjustment

By virtue of their work loading and material, these wires can become slack or stretched over time. The length of wire is altered by tightening or slackening the adjusting eyebolts on the hydraulic hoisting press. As each leg of the wire is to share an even load, pertinent points when making adjustments are:

- Periscope should be raised and on its pins.
- Crosshead clamps to be slackened.
- To avoid wires taking full diving pressure, the buffer on the bottom of the periscope to be just touching pad at bottom of well when fully lowered.

Periscope bearings

Periscope must not bump when raised. Wear in bearing ball races can cause vibration, so bearings must be greased regularly and tried by hand to see that they turn smoothly, checking that no radial slackness is apparent. Clearance should be 0.010in maximum and 0.004in minimum.

Jumper wire

Although not later fitted, any adjustment on this wire or inadvertent malpractice of slinging of loads from it can affect alignment of the periscope standard and consequently cause bumping of the upper periscope bearing.

Greasing

Greasing points to the three periscope bearings, crosshead ball race and the Blockhouse gland Plemil gun to be charged regularly (monthly).

Minor repairs

Due to the complexity of a periscope, only very minor repairs can be attempted by submarine staff. Full maintenance is to be undertaken by qualified technicians within dedicated ultra-clean periscope workshops, periscopes being fully removed in refit or as necessity demands.

Radar systems

Essential for communication, navigation, detection and surveillance, *Alliance* was fitted with various wireless telegraphy and radar systems. During her service each type was introduced according to the improvements made in this field. In general, all

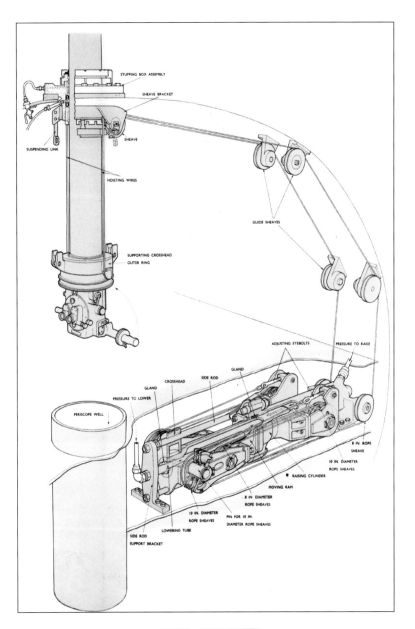

ABOVE Hoist gear for Type CK14 periscope.
(Plate 6, BR.1963/11)

LEFT Periscope hoisting cable and guide sheave.
(Jonathan Falconer)

ABOVE **Sonar display unit.** *(Jonathan Falconer)*

BELOW **Sonar dome at the bow.** *(Author)*

systems used antenna attached to dedicated telescopic masts.

All equipment employed comprised the following (numbers and letters given relate to their NATO Code designation):
- 267 MW (AVG fitted later) – mast located afore the attack periscope.
- ANF – mast located abaft the search periscope.
- 253 MW.
- 86 MW.
- 80 M.
- Whip aerial (fitted during the 1960s) – mast located starboard side of modified conning tower fin.

The wireless office was situated on the port side abaft the control room.

ASDIC or sonar systems

A submarine's 'hearing sense' while it is submerged comes from ASDIC (Anti-Submarine Detection Investigation Committee) or SONAR (SOund Navigation And Ranging). These systems work by means of a transmitter/receiver that sends out highly directional sound waves through the water. If a sound wave strikes an object, it is reflected back and the 'echo' is picked up by the receiver. The length of time from transmission to receiving the echo is used to measure the range, and the object that is detected is indicated as a flickering light on the range scale. With the transmitter head mounted so that it can be directed almost like a searchlight, target bearing can be read from the compass receiver.

The main Type 12A ASDIC is located at the foremost end of the keel, the Type 138B ASDIC under the casing abaft the engine room hatch, and the newer Type 183 sonar in its dome at the bow.

ASDIC or sonar can be used in two modes:
- **Passive** This is listening for the sound made by vessels.
- **Active** This is emitting pulses of sounds and listening for echoes; sonar may be used as a means of acoustic location and of measurement of the echo characteristics of 'targets' in the water. The disadvantage of 'going active' is that sound pulses emitted will give away the boat's position to the enemy.

Capstans and anchor gear

The capstan and cable holder are located on the exterior of the pressure hull above the torpedo tube space. Power is provided by a

10.5hp hydraulic motor supplied from the main telemotor system via a throttle valve fitted inside the boat on the deckhead of the fore torpedo stowage compartment. The actual valve that controls speed, direction of rotation and modes of hauling or veering is operated from the submarine casing.

Vertically mounted, the capstan is rotated by means of a series of worm drives and shafting that drive a combined four-snug cable holder and brake drum via a two-speed dog-clutched gearbox integrally connected to the two worm drives. The gearbox comprises a free sliding gearwheel that can either engage with its dogs with the fixed wheel to drive the cable holder by direct drive or mesh with a layshaft to give a faster drive ratio of 2.1:1. A cable compressor is fitted forward of the cable holder.

The cable holder, gearbox, brake band and telemotor control valve are operated through rod gearing with necessary expansion and universal couplings. A portable turn key is provided to operate the control gear.

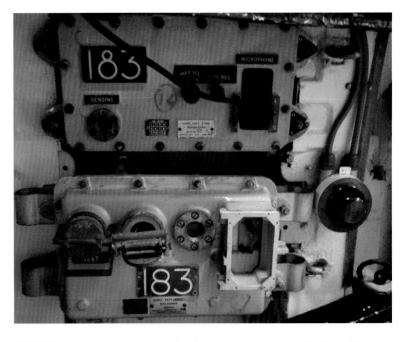

ABOVE Located in the fore torpedo stowage compartment, the Type 183 battery-operated emergency underwater telephone was used for escape. With a range of 1,000yd it can transmit for 12 hours and receive for 72. *(Author)*

BELOW Capstan and cable holder general arrangement. *(Plate 27, BR.1963/4)*

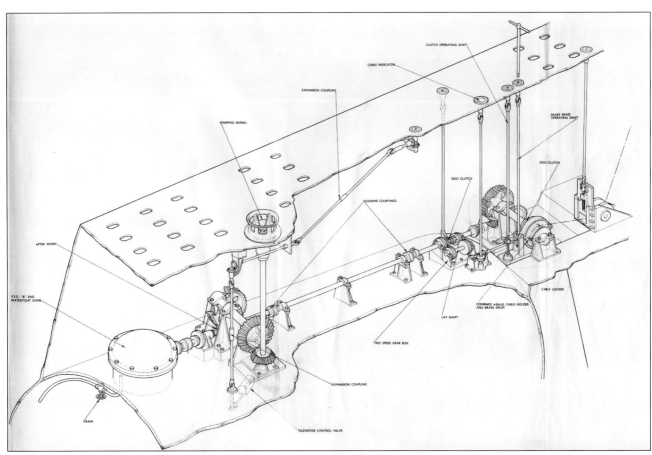

Chapter Seven

Operating *Alliance*

This chapter describes the procedures for operating HMS *Alliance* in the varied circumstances she encountered throughout her service career: diving, running submerged, 'running silent, running deep', rising to periscope depth, snorting, running surfaced, torpedo attack and 'diving stations'.

OPPOSITE Main blowing panel showing main ballast tank blow valves, main vent operating levers, and depth gauge. *(Jonathan Falconer)*

ABOVE **HMS *Alliance* at sea at speed.**

In all procedures described in this chapter verbal orders are given within quotation marks. It should be noted that submariners always use the word 'shut' (for valves, hatches, doors, etc) rather than 'closed', to avoid possible confusion with 'blows' (ie, the main or master blows), and in fact to use the word 'closed' is considered a mortal sin. Likewise, a crew member will 'make' switches and electrical breakers. Sources for instructions quoted below are based on SOPs (Submarine Operating Procedures) and author's experience and observations. Variations of procedure would also relate to SSOs (Ship's Standing Orders) drafted with additions, ie by the commanding officer, according to his preferences, and consequently procedures may differ from boat to boat, but the standard SOPs are still adhered to.

Diving the boat

1. 'Diving stations, diving stations, open up the boat for diving.'
2. 'Uncotter 1, 2, 3, 4 and 5 main vents.'
3. First lieutenant, coxswain and 'outside wrecker' walk through the boat making necessary compartment safety checks, all hatches shut and clipped; all hands not on watch close up at their dedicated diving stations.
4. 'Reduce speed to 4 knots.'
5. 'Shut down main engines.'
6. 'Shut group exhaust and muffler valves.'
7. 'Shut snort induction hull valve.'
8. 'Open numbers 1, 2, 3, 4 and 5 main vents' (main vent operating panel control room).
9. 'Open/shut (as required) O-comp outboard/inboard vent' (as required).
10. 'Out clutch; disengage diesel main engines.' Check clutch control indicators read 'clutch out'.
11. Confirm port/starboard group exhaust valves shut.
12. Confirm port/starboard muffler valves shut.

RIGHT **Main vent operating levers for No 3 main ballast tank (MBT) port and starboard, and No 4 MBT.** *(Author)*

13 'Group up and run main motor half ahead.'
14 Given bearing, helmsman brings boat head-on to sea.
15 Clear the bridge for diving – confirm conning tower hatches shut and clipped.
16 Flood O-comp accordingly by opening Kingston valves. 'Open/shut (as required) O-comp outboard/inboard vent' (as required).
17 'Diving now' (put planes to dive).
18 Bring the boat to 30ft and adjust trim using both sets of planes.
19 Adjust trim by running trim pump fore or aft as required.
20 'Cycle numbers 1, 2, 3, etc main vents' – confirm all are shut (this ensures air is expelled from tanks).
21 'Shut O-comp inboard vent.'
22 Bring the boat to periscope depth, raise periscope and make a visual 360-degree sweep.
23 'Down periscope.'
24 Group up or group down main motors, controlling speed on the field regulators as ordered on telegraph and take the boat to 250ft or depth as ordered and proceed.
25 'Stand down from diving stations, assume passage' (or patrol routine).

If diving the boat during snorting, the first order given is 'Stop snorting, stop snorting, stop snorting.' Diesel main engines are immediately shut down. Actions to shut off group exhaust valves, muffler valves and induction valves are carried out simultaneously, likewise the lowering of exhaust and induction masts. All other control actions related to diving the boat follow those given above.

Surfacing the boat

1 'Diving stations, diving stations, stand by to surface.'
2 Captain to the control room. Bring the boat to periscope depth, operating fore and after planes as required.
3 'Group down and run main motors slow ahead.'
4 If night-time, switch control room to red lighting (often termed 'revert to black lighting').
5 'Bridge watch keepers/special sea duty men close up.'

LEFT Starboard engine room telegraph (indicated green). *(Jonathan Falconer)*

6 Raise periscope (search or attack depending on operational limitations).
7 Make a visual 360-degree sweep on full power (to determine if ships, aircraft or other hazards are in the vicinity).
8 'Down periscope.' Bring helm to turn the boat's head to sea (surfacing with conning tower (fin) beam-on to heavy seas could incur excessive rolling).
9 'Up periscope' and make a final sweep. 'Down periscope.'
10 Check all main vents indicate shut.

BELOW Conning tower ladder leading up from the control room, with 'red' night lighting. *(Author)*

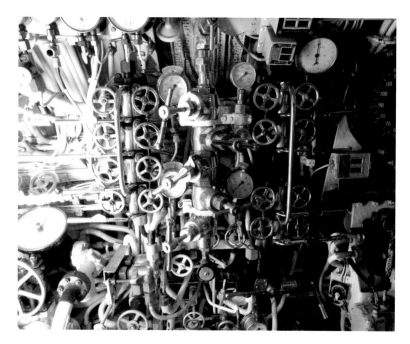

ABOVE Main blowing panel with master blow levers centre and main ballast tank main vent operating levers to left and right. *(Author)*

RIGHT Depth gauge. *(Jonathan Falconer)*

BELOW Steering (helm) position showing rudder control lever, gyro repeater, rudder position indicator and compass (above). Note clinometers to the right indicating degrees of roll on the boat. *(Jonathan Falconer)*

11 'Stand by to surface – blow numbers 1, 2, 3, 4 and 5 main ballast tanks' (blowing panel in control room).
12 Captain opens lower conning tower hatch and ascends conning tower, followed by bridge lookouts and Officer of the Watch (seawater normally descends into control room when hatch is opened).
13 First lieutenant assumes command in control room; watching depth gauge, he orders, 'Stop blowing, shut both master blows.'
14 Blow or pump O-comp tank as required.
15 'Start the low-pressure blower.'
16 Helmsman commences steering course orders via voice tube from bridge.
17 'Stop main motors, clutch in port and starboard main engines.'
18 When clutch indicator reads 'clutch-in', 'Start main engines.'
19 Open snort exhaust inboard and outboard drains; open group exhaust and muffler valves; blow exhaust mast and shut inboard and outboard drains.
20 'Stand down from diving stations, assume passage routine.'
21 Start high-pressure air compressors and commence recharging air bottle groups.
NB: If the boat is to remain on the surface for a sustained period (ie going into port) the commanding officer will give the order: 'Shut off from diving. Cotter 1, 2, 3, 4 and 5 main vents. Passage routine.'

Running the boat when submerged

This is mainly controlled by operation of the fore and after hydroplanes. The fore planes have a more significant effect while the after planes maintain level. Both planesmen pay particular attention to the position of the bubble in the clinometers. Trim is also maintained by means of the trim pump, which pumps water fore or aft between the fore and after trim tanks, altering bodily weight at the extremities of the submarine hull. Greater changes are made by means of the main line (ballast system) and the ballast pump transferring water from 'O' or 'R' compensating tanks as required. The Officer of the Watch is to be constantly aware of any changes that may affect the trim – such as

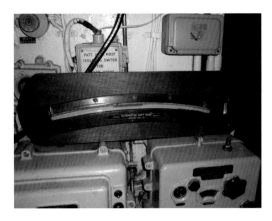

ABOVE Clinometer to indicate roll angle of the boat. *(Jonathan Falconer)*

RIGHT Port side hydroplane control positions; after planes to left, fore planes to right. Note depth gauges and bubble indicator clinometers; helm position on far right. *(Jonathan Falconer)*

transferring of oil fuel, discharging of sewage tanks or reloading of torpedoes – and make adjustments accordingly.

Keeping trim under all conditions (periscope depth, snorting or otherwise) is not only essential to good submarine practice, but constant operation of the trim pump and overcompensation of plane movement unnecessarily uses valuable battery power when the trim pump is run or when hydraulic plants cut in and out.

Speed is maintained by operating the main motor with the main batteries put in series or parallel, grouping up or down according to speed ordered and control of the field regulators; battery discharge and power levels available are constantly monitored.

Running silent, running deep

Operational requirements may dictate the avoidance of detection by an enemy vessel or when operating in covert conditions or making 'observations'. Precautions taken are as follows:
1 Take the boat deep using the varying layers of seawater density to screen detection.
2 Group down, run main motors slow ahead. This is done firstly to reduce shaft noise and consequently minimise detection by sonar and secondly to conserve battery power.
3 In order to reduce noise and conserve battery power, all non-essential machinery is to be shut down: for example, distiller, air-conditioning and refrigeration plants, lubricating oil separators and ventilation system fans. It may also be necessary to shut down one hydraulic plant. Lighting will also need to be reduced.
4 To restrict further noise throughout the boat, all 'off-watch' personnel are to 'turn in' in order to minimise unnecessary movement and vocal interaction.
5 Sonar operator will switch to passive and constantly monitor movement of surface attack vessels, reporting range, direction and speed as required.

Bringing the boat to periscope depth

Periscope depth for *Alliance* is approximately 54ft (16.5m) from the keel. If the submarine goes to periscope depth in order to carry out an attack, the boat will go to diving stations, with the entire crew and attack team closed up in a state of readiness. If periscope depth is used for simple observation or covert photography, the operation can be made on the standing watch.

In brief, going to periscope depth very much follows the actions and operations

ABOVE *Alliance* – masts and periscopes.

RIGHT Muirhead Mk 16 Torpedo Control Calculator (TCC) 'fruit machine' for computing torpedo attacks.
(Jonathan Falconer)

described in the section 'Running the boat when submerged' (above). The most critical difference, however, is that trim and depth are maintained to a very fine degree, without any erratic alterations. For this the coxswain will be called to take overall charge of both planesmen. Choice of periscope – search or attack – is dependent on operational needs, though use of the attack periscope is preferable to reduce the chance of visual detection by an enemy.

1 'Diving stations, diving stations. Standby to come to periscope depth.'
2 If night, switch control room to red lighting.
3 Officer of the Watch observes depth gauge. 'Planes hard to rise, bring the boat to periscope depth, Captain to control room.'
4 'Angle on the boat, start trim pump, pump X gallons' (fore or aft as required).
5 'Group down slow ahead main motors.'
6 'Up periscope.' Officer of the Watch of commanding officer makes an initial 360-degree sweep. 'Down periscope.'
7 'Up periscope'. Second sweep made to bring potential target/objective into vision.

The bearing (in degrees) at which the periscope is rotated and pointing at the objective is constantly repeated and recorded by the attack team member every time the periscope viewer shouts 'Bearing now.' This action is repeated as the submarine is manoeuvred under a succession of helm orders. This same information is constantly fed into the torpedo control calculator (TCC), colloquially called the 'fruit machine'. The TCC fitted in *Alliance* is of the Muirhead Mk 16 type developed during the Second World War.

If an attack is being made, the periscope will be constantly raised and lowered, minimising the time spent with it breaking surface. For similar reasons the boat may be taken down and brought back up to periscope depth throughout an attack, the entire operation requiring constant delicate adjustment and trimming by use of the planes and trim pump. The control room becomes a nerve centre with every man 'closed up' at his action station in the boat, acting independently but in total unison with the thoughts and actions of the boat's commander as an attack is made.

Snorting

The procedure for snorting very much follows the same actions and limitations as operating at periscope depth described above in relation to depth and speed keeping. As with use of the periscope, trim and depth have to be maintained very carefully to avoid the snort induction mast dipping below the sea surface and creating spasmodic vacuum in the boat. Regarding the snort exhaust system, this can discharge at a set depth below the surface provided that engine exhaust pressure remains sufficient to overcome sea pressure.

Ventilating the boat while snorting

This procedure may only be necessary simply to replenish the air in the boat when CO or CO2 levels are high, without running the diesels.

Procedure

1. 'Bring the boat to snorting depth.' Planesmen work in unison; trim pump run to maintain level in the boat. 'Outside wrecker', engine room artificer (ERA) and outside stoker to control room.
2. 'Revert ventilation system for snorting.'
3. 'Stand by to snort, raise the snort induction mast.'
4. 'Open snort drain hull valves 1 and 2 (in control room).' Outside stoker opens valves, watches water flow in tundish sight glass, shuts drain shut-off valve once water ceases to flow, and reports 'Snort induction mast drained; shut snort drain valves 1 and 2' (drain water passes to 'R' tank port).
5. 'Open snort induction line valve' (operated from engine room).
6. 'Open snort induction hull valve' (telemotor-operated).
7. 'Run ventilation induction fans.'

If simply ventilating when surface running, the alternative is to open the bridge induction line flap valve together with the bridge induction outboard drain. This avoids use of the snort induction mast, its associated snort induction hull valve and snort drains 1 and 2.

Charging main batteries while snorting

This procedure necessitates running the diesel main engines to drive the main motors as generators, producing DC current to charge the batteries.

Prerequisites

1. Both diesel main engines ready for running.
2. Both main engine turning gear clutches out.
3. Both main engine clutches out.
4. Air start reservoirs fully charged.
5. Boat brought to snorting depth.
6. Snort induction mast raised and drained as described above.
7. Battery fans running and ventilation system aligned accordingly.
8. Telemotor-operated snort induction hull valve and induction line flap valve open.

Procedure

1. 'Raise the snort exhaust mast.'
2. 'Stop both main motors.'
3. 'Engage main engine clutches.' Check clutch indicators are in.
4. 'Start both main engines.'
5. 'Engine room opens port/starboard snort muffler inboard drains and snort exhaust inboard and outboard drains.
6. Open group exhaust and muffler valves. As engine exhaust pressure raises, engine room artificer (ERA) blows the snort exhaust mast.
7. Set main engine revolutions as required for constant speed.
8. Shut snort exhaust inboard and outboard drains.
9. Motor room electrical artificer (EA) checks main motor field current at switchboard and makes switches to commence charging main batteries and checks battery charging panel.
10. Once charging, electrical rating to commence taking periodic specific gravity and temperature checks on battery electrolyte until charging is complete.

While the diesels are running, all stale air in the boat is consumed by the diesel combustion

LEFT Snort induction hull valve and vent valve indication panel with 'Q' tank contents gauge to the right. *(Author)*

ABOVE Sub-lieutenant on the bridge with the navigating compass.

RIGHT *Alliance* at sea, viewed from a following submarine.

process and replenished by air intake from the induction mast.

If charging when surface running, diesel exhaust can be directed aft through the surface exhaust muffler tanks by shutting muffler valves, exhaust passing out through the sides of the after casing. If running in this mode, seawater needed for cooling purposes is to be opened to the exhaust muffler tanks.

Running the boat when surfaced

With the diesel main engines acting as the main propulsion, the boat is operated in virtually the same manner as a naval surface vessel, with an officer of the watch, signal man, and communications and lookout ratings stationed on the conning tower bridge, with helm and engine orders relayed to the helmsman below. Vigilance must be paramount due to:

- The presence of other surface vessels, especially as a submarine presents a very low profile if running with the main ballast tanks partially filled and the casing awash. All submarines are particularly vulnerable to collision with other craft, especially at night, as evidenced by the loss of HMS *Truculent* when a Swedish oil tanker collided with her in the Thames estuary during the evening of 12 January 1950.
- Presence of enemy aircraft.

Diesel exhaust can be directed in two ways: overboard via the muffler tanks or via muffler valves to the snort exhaust mast. The latter mode gives the boat the ability to dive at very short notice and commence snorting.

Torpedo attack

The procedure given below applies to both Mk VIII and Mk XXIV torpedoes, but in the case of the latter, which are wire-guided, the weapon is guided from the torpedo guidance control system in the control room.

1. 'Diving stations attack team close up.' Standby for torpedo attack. 'Captain and First Lieutenant to control room.'
2. 'Shut torpedo room and torpedo stowage compartment bulkhead doors.'

RIGHT *Alliance*'s attack periscope.

3 'Bring boat to periscope depth' (see above for description of procedure).
4 'Flood up numbers 1, 2, 3, 4 etc (as desired) torpedo tubes (TTs)' (the torpedo tubes already being loaded).
5 Torpedo room reports 'Numbers 1, 2, 3, 4 etc TTs flooded up.'
6 Torpedo room equalises TT pressure and reports.
7 'Blow up numbers 1, 2, 3, 4 etc TTs.' Torpedo room reports when equalised.
8 'Open numbers 1, 2, 3, 4 etc bow caps.' Check indication 'open'.
9 'Group up or group down main motors half or slow ahead' (as required).
10 Sonar operator commences calling out target bearings and range.
11 'Up periscope.' Officer of the Watch or commanding officer makes an initial 360-degree sweep with the attack periscope.
12 'Down periscope.'
13 'Up periscope'. Second sweep made to bring potential target/objective into vision. The bearing (in degrees) at which the periscope is rotated and pointing at objective is constantly repeated and recorded by an attack team member every time the periscope viewer shouts 'Bearing now.' This action is repeated as the submarine is manoeuvred under a succession of helm orders. This same information, together with continuous reporting of target bearings and range by the sonar operator, is constantly fed into the torpedo control calculator (TCC).
14 'Down periscope.'
15 Firing control panel, directly linking control room and torpedo space (afore or abaft), is activated. From this point firing is directed from the control room, the actual physical action of operating the individual tube firing levers locally initiated within the tube space by torpedo ratings.

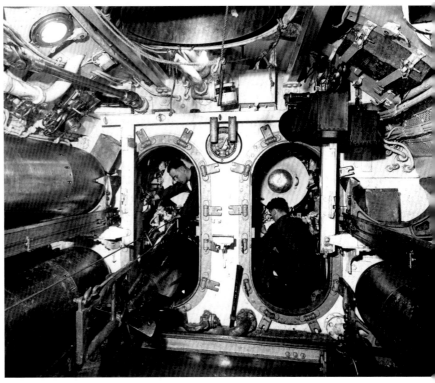

RIGHT In the forward torpedo compartment, showing a 21in torpedo being loaded into the torpedo tube. *(TopFoto)*

16 When torpedoes have been fired, bow cap doors are hydraulically shut when indicated 'shut', the torpedo tube is drained down into the torpedo operating tank (TOT) ready for opening the breech door for reloading (see Chapter 5).
17 'Stand down from diving stations, attack team stand down, patrol routine.'

Diving stations

This section covers procedures for surfacing the boat for gun action and shutting off when under attack from depth charges.

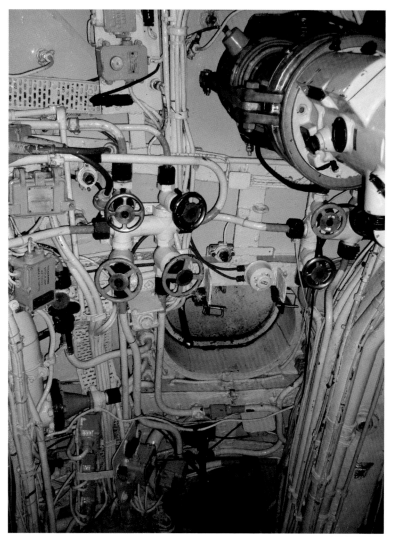

BELOW Inside the control room: overhead pipework and conning tower hatch with periscope to the right. *(Jonathan Falconer)*

Surfacing the boat for gun action

1 'Diving stations gun action attack team close up.'
2 Magazine keys passed to gunnery petty officer.
3 Wardroom vacated and ladder rigged to gun tower hatch. Gun's crew muster in wardroom; ratings as detailed on the watch and station bill form an ammunition chain between the magazine and wardroom and consequently up the gun tower.
4 Bring boat to periscope depth (as described above).
5 Captain and first lieutenant to the control room.
6 'Group down main motors to slow ahead' (there is no need to revert to diesel propulsion as the intention is to immediately dive on completion of gun action).
7 'Up search periscope'; Officer of the Watch (OOW) or Commanding Officer (CO) make an initial 360-degree sweep.
8 'Down periscope.'
9 'Stand by to surface blow 1, 2, 3, 4 and 5 main ballast tanks.'
10 Once partially surfaced, 'Shut both master blows and run the low-pressure blower.'
11 OOW and CO open the conning tower hatch and ascend to bridge.
12 Gun's crew open gun tower drain, open gun tower hatch and ascend to man deck gun.
13 Ammunition chain form up and commence passing shells.
14 Following orders given from command upon the bridge, gun's crew commence gun drill (see Chapter 5).
15 On completion, gun is prepared for diving, gun's crew and ammunition train descend into boat, shutting and clipping gun tower hatch.
16 OOW orders 'Diving stations' and vacates the bridge with bridge party, descends to the control room, shutting and clipping conning tower hatches.
17 'Open main vents.' 'Diving now.' 'Bring the boat to X feet' (depth as ordered). 'Flood Q.' Flooding 'Q' tank may be necessary to dive the boat more quickly to avoid possible visual detection.

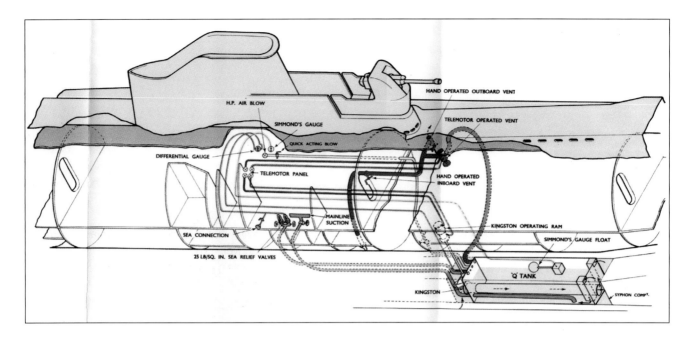

ABOVE **'Q' Tank and its associated services.** *(Plate 4/2, BR.4549/BR.2507/2)*

Shutting off for depth charges

When under attack the immediate action is to take the boat deep to avoid depth-charge explosions crushing the hull or compromising hull integrity, and to use the change in water density to help to reduce the possibility of detection by sonar.

1. 'Diving stations. Shut off for depth charges.' All compartments manned by off-watch personnel, captain and first lieutenant to control room, engineer to engine room.
2. 'Take the boat to 250ft' (76m). Planes put hard to dive, angle on the boat as ordered.
3. 'Shut all bulkhead doors.'
4. 'Shut down all bulkheads'. Ventilation and non-essential system bulkhead valves etc shut and isolated.
5. 'Man all compartment blows.'
6. 'Group down, run main motors slow ahead'. This is done for two reasons: to reduce shaft noise (and minimise detection by sonar) and to conserve battery power.
7. 'Shut down all non-essential machinery.' Again, this is done to reduce noise and conserve battery power; machinery affected includes distiller, air-conditioning and refrigeration plants, lubricating oil separators and ventilation system fans. It may also be necessary to shut down one hydraulic plant.
8. 'Reduce lighting throughout the boat and rig emergency torches.' Lighting breaker switches made will cause compartment action emergency lighting (AEL) to automatically come on; this lighting is sustained by an integral battery supply.
9. As a precautionary measure, minor valves and cocks on all systems not tested to deep diving depth (DDD) are shut. The boat is now in a high state of readiness to resist potential attack and damage; running silent and deep.

When running silent and deep, the sonar operator will switch to 'passive' and constantly monitor movement of surface attack vessels, reporting range, direction and speed as required. During this time it may be necessary to take the boat deeper, close to DDD, to avoid damage, and it may also be imperative to constantly alter course in order to confuse a pursuer.

Despite the nervous tension enforced upon the crew members in this critical situation, absolute vigilance is applied by all personnel in identifying problems – such as any breach to system valves or pipe flanges – and then isolating systems, undertaking repairs and making safe electrical supplies as necessity dictates. Each compartment makes damage reports by phone to the control room in order for the CO, first lieutenant and OOW to assess damage sustained and the operational condition of the boat.

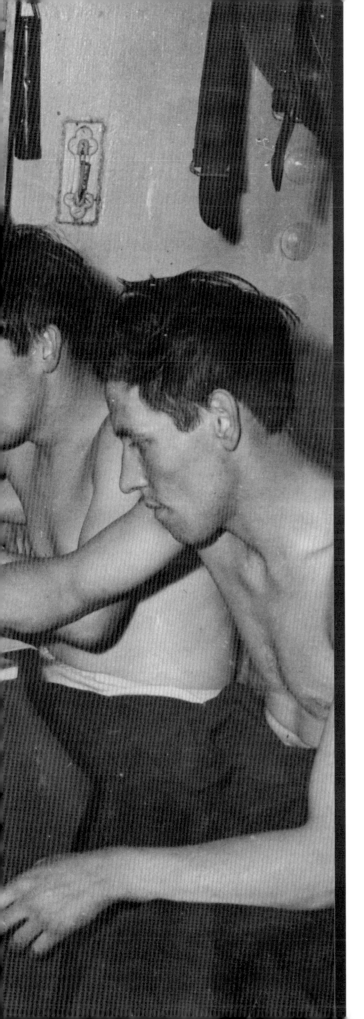

Chapter Eight

Manning *Alliance*

Submariners are a unique breed. All crew of a 'boat' share extraordinary camaraderie while retaining the ability to keep their individuality in confined conditions. This chapter describes the crew and their duties, and presents reminiscences about living conditions from those who served on A-class submarines.

OPPOSITE Off-duty crew members relax in their mess onboard HMS *Alliance*.

The crew and their duties

Men serving in the submarines had to be volunteers and over 18½ years of age before being allowed to enter the service. Junior ratings, therefore, had to be able seamen or of first-class rate within their branch specialism, and all would be recognised for their good conduct.

Training for the Submarine Service

Despite previous branch specialism, all candidates completed their initial Parts I and II training at HMS *Dolphin* Submarine School. This comprehensive course covered all matters of submarine systems machinery, tanks and their operation, and all aspects of safety. After sitting respective examinations and successfully passing escape training at the submarine escape training tank (SETT), candidates would continue their Part III training on the boat to which they were drafted.

Part III comprised thorough on-board training related to system identification, operation and ship safety, during which the trainee, colloquially called a 'Part III', would have to demonstrate his knowledge and competence to respective crew members and divisional superiors, and have his 'Part III book' (in which he would have drawn system diagrams and completed questionnaires) inspected and signed off accordingly. This applied to all submarine personnel – even cooks had to have basic submarine engineering knowledge. The candidate would then sit a viva voce (verbal exam) with his head of department (HOD) senior rating and his appointed divisional officer (DO).

Failure to succeed within three months could mean loss of submarine pay or, worse, dismissal to serve back in the 'Gens' (Royal Navy General Service surface fleet), a fate considered abominable.

The ship's company

Alliance's complement numbered 63 men, comprising 6 officers, 16 senior ratings – non-commissioned officers (NCOs) – and 41 junior ratings as outlined in the accompanying table. They were formed into five divisions (departmental or branch specialists), the largest groups – the Seamen and Marine Engineering departments – forming 72 per cent of the crew.

BELOW Crew members on the bridge of *Alliance* at sea off the coast of Portugal. The boat is in her as-built condition.

CREW

Seamen branch (5 officers, 3 senior ratings, 15 junior ratings)		
Rate	Title	Number
–	Commanding Officer	1
–	Executive Officer	1
–	Seaman Officers	3
Petty Officer	Petty Officer Seaman, the senior designated as the Coxswain	3
LS	Leading Seaman	3[1]
QA2	Quarter Armourer	1
UW1	Under Water Weapons	1
UW (star)	Under Water Weapons	4
UC1	Under Water Control	1
UC2	Under Water Control	2
RP2	Radar Plot (also acts as Navigator's Yeoman)	1
RP (star)	Radar Plot (also acts as Navigator's Yeoman)	1
Sub-total		**22**
Communications branch (1 senior rating, 4 junior ratings)		
CRS	Chief Radar Supervisor	1
RS	Radar Supervisor	1
LRO (G)	Leading Radio Operator	1
RO (G)	Radio Operator	3
Sub-total		**6**
Marine engineering (engine room) branch (1 officer, 8 senior ratings, 16 junior ratings)		
–	Engineer (Officer)	1
CEA (P)	Chief Engineering Artificer	1
MEA (P)	Marine Engineering Artificer	4
Mech 2 (P)	Mechanician (Mechanical Engineering)	1
CMEM	Chief Marine Engineering Mechanic	1
POMEM	Petty Officer Marine Engineering Mechanic	1
LMEM	Leading Marine Engineering Mechanic	4
MEM	Marine Engineering Mechanic	12[2]
Sub-total		**25**
Weapons and electrical engineering branch (3 senior ratings, 4 junior ratings)		
POREL	Petty Officer Radio Electrician	1
CEA or CEMn	Chief Electrical Artificer or Chief Electrical Mechanician	1
POEL	Petty Officer Radio Electrician	1
COEA or COE Mn	Chief Ordnance Electrical Artificer or Chief Ordnance Electrical Mechanician	0[3]
OEA or OEM	Ordnance Electrical Artificer or Ordnance Electrical Mechanician	1
LOEM	Leading Ordnance Electrical Mechanic	1
OEM1	Ordnance Electrical Mechanic First Class	2
Sub-total		**7**
Supply and secretariat (1 senior rating, 2 junior ratings)		
LCK	Leading Cook	1
CK	Cook	1
LSTD	Leading Steward	1
Sub-total		**3**
Total		**63**

Source Admiralty Letter N/M135/56 dated 10 February 1956

[1] One extra LS carried if 4in gun fitted, to act in QA2 duties.
[2] One extra MEM borne in war.
[3] One COEA or COE Mn per squadron.

ABOVE Captain and Executive Officer with ships' company in their mess.

Commissioned officers

There were six commissioned officers as follows:

RIGHT ERA Bill Handyside.

- **Commanding Officer (CO)** In overall command, this man held the rank of lieutenant-commander; commanders of submarines have always been of lesser rank than those serving in the surface fleet despite their greater responsibilities. As seen from the service history of *Alliance* outlined in Chapter 2, command was occasionally taken by a lieutenant.
- **Executive Officer (XO)** Second in command, this officer held the rank of lieutenant.
- **Seaman officers** Forming the rest of the executive officers, these men generally consisted of a first lieutenant (navigator), a lieutenant and a sub-lieutenant (or sometimes two sub-lieutenants) undertaking watchkeeping duties in the control room.
- **Engineer** Holding the rank of lieutenant, this man was fully responsible for propulsion, electrical systems, auxiliary machinery hull systems and equipment. Trained in the disciplines of either E (Electrical) or M (Mechanical), his combined abilities covered both skills. Unlike the other officers, he did not generally stand any watches.

LEFT 'Tot time' – PO Thompson and Leading Seaman Forster.

Key senior ratings

The other key personnel were as follows:

- **Coxswain** Colloquially addressed as 'Swain'. As well as being the chief helmsman who would close up at the helm at diving stations, he had numerous other duties. He regulated the ship's company with respect to discipline and berthing the men. He acted as supply officer, provisioning the boat with stores. He was in charge of victualling and responsible for embarking foodstuffs, a role that gave him the nickname 'Grocer', and he also oversaw the daily issue of rum. Sufficiently trained in first aid, his role as the 'medic' meant he could deal with broken limbs, stitch wounds and general sickness.
- **Chief MEA (P)** Colloquially called the 'Chief ERA' (Engine Room Artificer), this man is directly responsible to the Engineering Officer for all matters of machinery operation and maintenance, including the hull systems. Very much a 'hands-on' man, his initial training as a precision fitter and turner meant he could fix virtually all machinery, and could even manufacture new components at sea if necessary. The suffix (P) relates to propulsion
- **CEA (P) Chief Electrical Artificer** directly responsible to the Engineering Officer for all matters of operation and maintenance of heavy electrical machinery battery, main motors and power systems. The suffix (P) relates to propulsion.
- **Torpedo Anti-Submarine Instructor (TASI)** The senior petty officer in charge of all matters concerning torpedoes.

BELOW Crew on the bridge.

■ **Chief** Marine Engineering Mechanic Usually referred to as the 'Chief Stoker', he was responsible for fuelling and fresh water, and kept alternative watches to the 'outside wrecker'. He was also divisional officer to the stokers (marine engineering mechanics).

Medical staff

A-class boats rarely carried any medical ratings except for operational squadron duties, in which case one man was allocated per squadron. In these instances the boat would carry either one LMA (Leading Medical Assistant) or one MT4 (Medical Technician). Under normal circumstances the Coxswain dealt with all medical matters, including minor surgery.

Living in the boat

A submariner has 'attitude'. A submariner, his boat and his crew always come first. The British submariner also holds a healthy contempt for his counterparts in 'Gens' (General Service), the Royal Navy's surface fleet, known as 'skimmers'. Conversely, he will hold all other submariners, of whatever nationality, in high esteem – a 'band of brothers' unified by common conditioning in an alien, hostile environment.

Despite their unconventional 'piratical' attitude to the world in general, submariners are intolerant of injustice. Rigorously professional, the submariner is 'careful but carefree'. He expects and accepts working long hours, tolerates cramped, crowded conditions and compromised personal hygiene, and endures the permeating smell of diesel, shale oil, cooking and body odours.

Life in *Alliance* is best summed up by reminiscences of her crew gathered together in the rest of this chapter under a variety of themed headings.

Cold War patrols during the 1960s

Operating quietly when submerged, *Alliance* would be covertly deployed on 'sneakies'. Alan Baker recalls: 'A lot of the time you really didn't know what *Alliance* was doing day to day. When you were on a "mizzy" [mystery tour] you ceased to exist. Your wife could never find out where you were but knew you were doing a "mizzy" if the pennant number was painted out. You were never told how long you were going for; it was six to eight weeks but you were never told exactly.'

Working as an Engine Room Artificer (ERA)

Drafted into submarines in 1953, Bill Handyside preferred working in submarines because self-discipline was greater than in battleships, where he found life too regimented: 'You more or less had your own job to do and there was less being directed ... in submarines you concentrated on the work and keeping the boat running.'

Working in the engine room, Bill kept watches one hour on and four off, standing at the engine control panel when the diesels were running: 'The noise was terrific, no ear plugs but you would get used to it. Listening was important ... you could tell if anything was wrong when the noise of the engine changed.'

Bill recalled when *Alliance* travelled to Canada: 'We went full-tilt for seven or eight days, watch-keeping two hours on two hours off, never stopping at all until the end of the journey.' On the way to Quebec one of the diesels sprung a leak from a cylinder and the whole unit had to be changed, involving the use of heavy spanners

BELOW *Alliance* at sea with crew members relaxing on the forward casing.

and chain blocks as lifting gear. 'We completed this in 18 hours – the fastest it had been done. Working like that in terrible conditions makes you bond – no shirkers.'

Regarding workshop facilities Bill said: 'On *Alliance* you had a small 4in lathe so for any small jobs you carried the raw materials under the engine room plates. Any small jobs you could do yourself and all of us were tradesmen, mostly fitters and turners, but if you were a boilermaker or coppersmith it didn't make any difference. The thing I remember about it was the skill they had in making things and it was terrific how you managed to keep going. I'm sure that other navies couldn't keep going like we did. If something broke down you'd do your best to repair it … with the skill and inventiveness of the ERAs you would always keep the engines running one way or another, or other machinery.

'You are overcrowded on submarines, living and working on top of each other. But there's camaraderie; in the mess there would be five or six of you and you could leave your watch and wallet out, and it would still be there later. There was feeling towards each other and doing little things for each other like covering someone's duty. You stay friends long after you leave the service; I've had one friend for 50 years.'

Cooking

Cooking for some 60 men in a confined galley was no mean feat. Bill Handyside recalls: 'Food compared to general service was pretty good. The chefs performed miracles with what they had. How two chefs, working in a very tiny galley in terrible conditions, produced three cooked meals a day for 60 men … well, I don't know how they did it.

'A favourite was "babys' heads" [steak and kidney pudding]. We had a lot of chips and gravy, steak sometimes, sometimes fish. Sometimes we would come across a fishing boat in the Channel and they would give you fish and crabs. After a while the bread would go mouldy, so we would cut the mould off until all the green had gone; but at times the chefs would bake their own bread.'

Leading Cook Kevin 'Pony' Moore served in various A-class boats during the 1960s and '70s: 'It was no picnic. You were always the last person to be told from the officer in the control room what the boat was about to do. Frying eggs when running on the surface could be a disaster. As for baking bread, just as your loaf was starting to rise the OOW [Officer of the Watch] would suddenly take the boat up to periscope depth to start snorting, and the vacuum drawn through the boat by the diesels would suddenly deflate your loaf into a doughy mess in the oven and you had to start all over again. Sometimes the OOW would start doing 'angles and dangles' [turning, diving and rising the boat through various depths] and you never got told. Soup, beans and custard went everywhere, usually over you as well as the galley deck, likewise your stew or spuds. Despite the difficulties, we still always provided hot meals as it was far better for morale.'

Note: 'Pony' Moore served in *Acheron*, in which the author also served briefly.

Food storage and preparation

When provisioning a submarine for sea, food would be stored wherever space could be found. Bread was stowed on slats of wood laid on the torpedo-loading rails, and vegetables stowed in the 'cabbage patch' in the fore ends.

Jim Onions (1950s): 'The Coxswain had a rum store and dry provisions store at the after end

LEFT Chef at work in the galley.

of the battery and he would put as much in his store as possible, but the rest was spread out throughout the submarine. Each mess took their potatoes etc to the galley with a note saying which mess they were from. Meals were always on the change of a watch. Boxes of tins would create walkways in the fore ends, the accommodation space and the after ends, but never in the engine room, because of contamination, or the control room, because they would get in the way of the attack team. The longer you were at sea the more headroom you got!'

Mick Daish (1960s): 'The junior rates' wash place was opposite the galley but you could never wash your face as the sink was always full of spuds! When you were in the engine room or fore ends on spud-peeling duty, you would be peeling for the next meal but one; all the spuds would be black because of the dirty hands! You could still see the oil marks on the spuds when they were roasted.'

Peter Pierce (1969–71): 'They used to store UHT milk in the shower cubicles on the *Alliance*!'

Mealtimes

Mealtimes were an important social gathering of the day and good for morale, especially when the rum tot was issued. Records indicate that half a pound of meat per man per day was supplied.

Hugh Ross (1940s): 'Someone would shout that the meal was ready and that was it. How was it dished up? In earlier boats it was on trays and you dug in from the floor, but on *Alliance* it was better … it was put out on the table and the senior man would dish it out in case someone got a bit greedy. It was shared out fairly. We fed well on *Alliance*.'

Bill Handyside (1950s): 'Food had to be carried from the galley forward through the control room to the cramped messes where it was eaten. In bad weather it was difficult and it would upset the trays of food, and at night when you went through the control room in red light, you would bump into people.' Author's note: The control room was always under red lighting at night to assist night vision when going up on to the conning tower or looking through the periscope; I recollect that red lighting always gave a macabre appearance to people in the control room.

Mick Daish (1960s): 'Dinner was at 12, tea was at 4 – cheese and biscuits. Supper was at 6. Then there was 9 o'clockers, dried stuff, you helped yourself.'

Lewis Whittaker (1973): 'One of the best meals was while we were in Campbeltown tied up to a buoy. At night time they caught 69 mackerel and the chef cooked them for breakfast.'

When loading missiles at Coulport, the 'off-watch' stokers often fished and supplied the galley. One time a great conger eel was vigorously fought with on the fore casing by stokers and finally dispatched by the 'trot sentry' – whose job was to stand watch on the casing – with his defensive pickaxe handle.

Sleeping berths

Bill Handyside (1950s): 'In the ERAs' mess we all had a bunk of our own whereas in the ratings' messes they used to have to "hot bunk", which meant that when somebody vacated a bunk to go on watch somebody else coming off watch would use it. We couldn't sit down in the mess until we'd folded the bunks away, then you're on top of each other sitting round the table most of the time. There was no room to walk around at all; you either sat down on the lower bunk or round the table or you pulled the bunks down and went to sleep.'

Jim Onions (1950s): 'I first slept in a hammock across the stokers' mess as there were too many people onboard for the number of bunks, but I got a bunk after a couple of months. You couldn't turn over in the bunks; they were canvas and you didn't have a mattress and pillow. This was before modernisation, different to what is in *Alliance* now.'

Mick Daish (1960s): 'If you were last on the boat you got a coffin bunk, one of the bottom ones, plus you hot-bunked, according to age and seniority.' Daish was temporarily seconded to the *Amphion* in Singapore and didn't have a bunk because of the extra Canadian crew the boat was carrying, so a camp bed was made up for him and placed right next to the torpedoes, one of which broke free in rough weather.

Alan Baker (1960s): 'You didn't get undressed to sleep except for your wet steaming boots. I slept underneath a stoker who had very smelly feet. A bunk was 5ft 10in long, so if you were taller than that you had

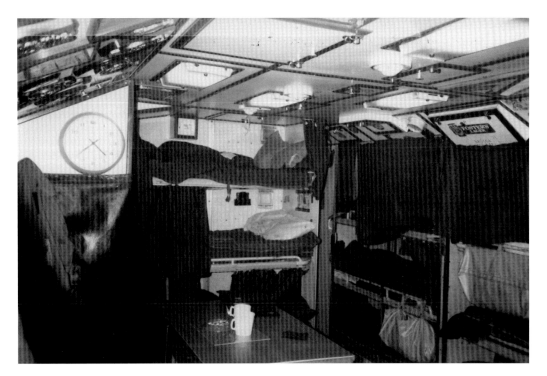

LEFT Mess, showing bunks.

to sleep slightly curled up … you had a bit of privacy when you drew the curtains on your bunk but there was still the noise; you could never shut out the sound but you learned to sleep through it. Everything was transparent, everyone knew your business.'

Lewis Whittaker (1973), a lieutenant who had a bunk in the wardroom, recalls the captain's use of the wardroom: 'Used to try and throw him out but he wouldn't go … He liked to be in the wardroom as it was the nearest place to the control room. In that boat the captain's cabin was halfway up the conning tower, which isn't the easiest place to get to, especially in an emergency.'

Clothing and cleanliness

Bill Handyside (1950s): 'As an ERA I wore the same clothes all the time. I wore my overalls in the engine room, I wore them in the mess and I wore them at night in the bunk when I was asleep, with my boots on – you never took those off. I was ready for anything, day and night, in case you had to go to diving stations. In ten days on patrol we never washed clothes except maybe a pair of socks now and then. You stayed a bit scruffy and you used to smell a bit I suppose, but you waited until you got back to harbour. If you had been doing something dirty, you would use diesel fuel to clean yourself. You put a bit of water in the bottom of the basin and cleaned hands and face, then went back to mess. Water was always short; although the *Alliance* had distillers for water, they were always going wrong.'

Note: Bill Handyside is now a volunteer guide for the *Alliance*.

BELOW Wardroom looking aft.

Environment, health and medical arrangements

As *Alliance* did not carry a trained doctor, all medical needs were undertaken by the coxswain, who would have received elementary training in dealing with general illnesses, burns, stitching up wounds, setting fractures and extracting teeth. The coxswain kept a medical book for guidance. Anyone suffering a debilitating illness was confined to the 'sick berth', a dedicated bunk located on the starboard after corner of the fore torpedo compartment. If on covert operations the crew simply dealt with any intervening medical situation without aborting the patrol.

Bill Handyside (1950s): 'After diving the air got a bit muggy. When you surfaced again and you got fresh air it was great. The Captain would order "One all round" – you could have a cigarette. Everyone smoked in the submarine, but you were not allowed to smoke when charging batteries because hydrogen is given off and it's too dangerous. When on passage routine all vents were cottered to prevent them opening. But I was never afraid, never claustrophobic, and would welcome being dived as the boat became stable. I was never unhappy about going to sea in a submarine.'

Alan Baker (1960s): 'The only place you had privacy was the toilet. You learned to get along with the people you didn't really like or had a connection with. You remember the cold, especially the huge force of air coming through the conning tower, chilling the whole boat. There were some [wall-mounted] heaters but half of them didn't work. Often lights and heating were turned off as it was thought to be a drain on energy. Stale air gave you a headache, the paracetamol made you constipated. I can't recollect anyone having a cold, though a cold wouldn't survive on the boat!'

Author's note: One of my submarine experiences was of a motor room watchkeeper developing tuberculosis, causing all crew to be medically screened for one year.

Recreation

Bill Handyside (1950s): 'When you came off watch you washed your hands and face, went into the mess and lay on your bunk or played cards. In *Alliance* we ERAs mainly played bridge or read books.'

RIGHT Crew members at mess table in the 1960s.

Alan Baker (1960s): 'The mess on boats was small and had bunks which folded down to make the seats. So you only had a small window of time to spend in there and you often couldn't do anything that made a noise so you read or played cards. Once a week we watched a film.'

Mick Daish (1960s): 'At sea you were either on watch, eating or in the bunk. There wasn't too much of a social gathering in an evening, maybe the odd game of 'uckers' [a form of ludo] and occasionally you would show a movie with a sheet hung up by the torpedo tubes as a screen; Royal Navy films were reasonably current ...'

Gordon Coles (1960s): 'News sheets came round but these were copies of what came out of the WT [wireless telegraphy] office. A couple of people took up knitting! There was also a Scalextric set rigged in the sailors' mess. One person had a guitar, but it was smashed by someone called Edwards because it got on his nerves.'

Bill Handyside (1950s): 'You would get sippers or gulpers of rum. When rum came up each day, the chief ERA would put a bit of water in it – enough to go round again. You would watch everyone to see if they were sipping too much.'

Vermin

Many boats suffered infestations of vermin of one kind or another, most often cockroaches but rodents were also common.

Mike Draper (1950s): 'There was a huge problem with cockroaches; people would use jars with coffee at the bottom and butter round the rim to catch them. There was also a problem with rats and in dock everyone had to leave the boat for it to be fumigated; dead rats, horrible smell, would be found a week later.'

Ron Potts (1960s): 'It would take a weekend fumigating the boat, and then we would have to shovel up the cockroaches when we went back on. Because they were laying eggs it was an ongoing cycle; as soon as you had got rid of one lot, the next lots of eggs would hatch.'

Author's note: I recollect that flies quickly died after the start of a patrol.

ABOVE Crew members in their mess at tot time, November 1964.

LEFT WT office – the telegraphist is taking a message.

Chapter Nine

Submarine escape

Successful escape from a submarine involves relatively sophisticated equipment together with effective evacuation procedures. Individual escape from a sunken submarine of *Alliance*'s era and type was the accepted method of saving life.

OPPOSITE HMS *Thetis* **rescue attempt, June 1939.**

ABOVE HMS *Affray* memorial at Gosport esplanade erected by the *Affray* Submarine Memorial Trust in April 2013. *(Author)*

The first successful escape was made from the German submersible *Brandtaucher*, which sank in Kiel harbour in 1851 after a trim weight slid forward, causing the boat to dive uncontrollably and hit the seabed at 60ft (18.3m). As water seeped in, Wilhelm Bauer, *Brandtaucher*'s designer, immediately understood how to resolve the predicament in which he and his two crewmen found themselves. Fighting against natural instinct, he forced his crewmen to wait with him for six and a half hours until pressure in the boat equalised with external sea pressure, Bauer himself admitting water at intervals to facilitate this. Finally opening the hatch and using the air bubble created, the three men shot to the surface intact.

As submarines developed, escape methods improved and reduced the margins of risk, life being preserved at the expense of the submarine. Today's submarines are adapted to use the Deep Sea Rescue Vehicle (DSRV), which not only enables the safe removal of men but also potentially permits submarine salvage.

Submarine loss

Compared with other maritime forces, the safety record for Britain's Royal Navy is second to none, with only two serious submarine incidents involving fatalities since the Second World War. Ignoring wartime conditions, the three factors that cause loss of a submarine – a highly abnormal occurrence – are given in the following sections, together with historic examples.

Total or partial mechanical failure

HMS *Affray*, 16 April 1951 A sister submarine to *Alliance*, *Affray* was lost in the English Channel, on the northern edge of Hurd's Deep, with the loss of 75 lives (see Chapter 2). The probable cause was a defective snort mast induction valve.

HMS *K13*, 19 January 1917 This K-class submarine sank in Gareloch, Argyll, Scotland, during sea trials when an intake failed to close while diving, causing her engine room to flood. Her loss was only noticed when a crewman surfaced after escaping. Eventually salvaged, *K13* was recommissioned as *K22* in March 1917.

Design failure

HMS *Thetis*, 1 June 1939 This T-class submarine sunk with the loss of 99 lives in Liverpool Bay during trials while undertaking its first dive. The cause was inadvertent opening of the inner torpedo tube door while the bow cap door was open as a result of two design failures: first, a test cock failed to indicate that the tube was full of water and, second, the bow cap indicators were arranged in a confusing configuration. As a result of this accident a modification was introduced: called a 'Thetis clip', this is a latch that permits a torpedo tube door to be opened only a small amount in case it is open to the sea at the bow cap door

RIGHT HMS *Thetis* following salvage after she had sunk during an accident off Holyhead on 1 June 1939, with the loss of 99 lives. *(TopFoto/The National Archives)*

end. Once it is clear that no flooding will occur, the latch can be released and the door fully opened. Salvaged and repaired, *Thetis* was recommissioned as HMS *Thunderbolt*.

Human error
HMS *Truculent*, 12 January 1950

Submarines are at their most vulnerable when operating in shallow waters – river estuaries, harbours and coastal waters – where the risks of running aground and collision are greater. When running on the surface, especially at night, a submarine is not obviously visible to surface vessels by virtue of its low profile and therefore it is vulnerable to collision. The classic example of this is the accidental sinking of HMS *Truculent* in the Thames estuary after a collision with the Swedish merchant ship *Divina* while returning to Sheerness after a refit at the Chatham Dockyard. After a premature escape attempt 57 of her crew were swept away in the current, with 15 survivors rescued by a boat from the *Divina* and five by the Dutch ship *Almdijk*. Salvaged on 14 March 1950, *Truculent* was beached at Cheney Spit for investigation. A newspaper report from the *Sydney Morning Herald* dated 18 March 1950 read:

Experts who are studying the beached wreckage of the submarine Truculent *believe they have discovered one reason for the magnitude of the disaster. When the Swedish ship* Divina *and the* Truculent *collided on January 12, 63 members of the crew lost their lives. The experts found that at the entrance to the forward torpedo-room a watertight door had jammed three inches open. They believe that, if the door had closed, 'the submarine might have remained afloat'. The Admiralty expects that, behind the door, divers will find the bodies of 16 men who were trapped as they were having supper. Divers so far have recovered two bodies from the forward control room. There may be a total of 20 bodies in the hull.*

HMS *K4* and *K17*, 31 January 1918

These two K-class submarines were lost while exercising with the fleet off the Firth of Forth as a result of collision, as happened with HMS *Truculent*. Later colloquially called the Battle of May Island, this incident commenced when the cruiser HMS *Fearless* collided with *K17*, which was heading a line of submarines. As *K17* sank, in just eight minutes, other submarines following her turned to avoid the cruiser. In so doing, *K4* was struck twice, first by *K6* (which almost cut her in half) and then by *K7*, and sank with all her crew. At the same time *K22* (the salvaged and re-commissioned *K13*) collided with *K14* but both survived. Within 75 minutes 2 submarines had been sunk, 3 had been badly damaged and 105 submariners had been killed. As an aside, *K4* had collided with *K1* off the Danish coast the previous year, *K1* being scuttled to prevent capture.

HMS *Artemis*, 1 July 1971

This A-class submarine sank in nine metres of water while moored in Haslar Creek, Gosport, as a result of a succession of minor procedural failings during refuelling that led to disaster.

'Sub-miss' and 'Sub-sunk'

Just before a submarine dives it generally provides a radio signal (operational requirements permitting) to home base after which time radio silence is maintained until its return to the surface on completion of its mission. Should no signal be received at the time expected, the naval authorities commence a procedure called 'Sub-miss', during which period of time surface ships and other submarines operating near to the assumed location of intended surfacing are alerted that a potential problem is imminent.

If no signal has been received after a period of 24 hours has elapsed, it is then assumed that the submarine has met difficulties, in a dangerous situation or sunk. At this point the Royal Navy immediately initiates a procedure called 'Sub-sunk', when surface ships and other submarines operating nearby are directed to commence a search using visual and sonar means to identify the position of the boat. If a submarine indicator buoy has been released, been seen, and its distress signal been received, Sub-sunk procedure is immediately initiated.

Submarine escape training tower (SETT)

Run by the Royal Navy for training submariners in the various methods of emergency escape from a disabled submarine,

ABOVE Submarine escape training tower (SETT) at Gosport viewed from the *Alliance*. (Author)

BELOW Escape training inside the SETT tank at Gosport. (Courtesy yorkshire-divers.com)

the submarine escape training tower (SETT) is located at Fort Blockhouse, not far from the Royal Navy Submarine Museum.

This facility comprises a 100ft (30m) deep cylinder filled with water with several entrances at varying depths, each simulating an air lock in a submarine, its bottom part being formed as a real part of the fore ends of a submarine. A team of experienced diving specialists called 'swim boys' man the tank, supported by a medical team with a necessary decompression unit. Besides physically escaping 'free ascent' from depths of 30ft (10m), 60ft (20m) and 100ft (30m), trainees also make 'suit' escapes from 100ft.

The highly comprehensive training is underpinned by lectures and practical tuition in how to survive within a disabled submarine, operation of emergency equipment and survival techniques on reaching the surface, the whole being a package of potentially lifesaving skills. Because escape involves personnel being subjected to relatively high seawater pressure and the possible risk, like divers, of suffering the 'bends', all submarine candidates are given a severe medical examination for physical fitness beforehand. Any deficiencies – such as insufficient lung capacity or heart problems – bar entry into the service, and no man can become a submariner until successfully passing the escape course. Every three years each submariner is recalled to requalify, albeit those aged over 40 are exempt to avoid unnecessary risk.

The SETT at Fort Blockhouse (HMS *Dolphin*) has been in use since 1954 so most crew members of *Alliance* would have undergone escape training there. It has also been used to train submariners from many foreign navies, including Italy, USA, Greece, Canada, Israel, Russia, Venezuela, Turkey, Australia and the Netherlands. The staff and facility have a worldwide reputation for excellence and good practice.

Author's note: During my training, when making a 'free ascent' from 100ft, I commenced blowing out air hard as instructed. After rising just 40ft (12m) I was short of air and fleetingly feared that I did not have enough for the remaining 60ft (20m) to the top of the tank. However, rising a little further, the little air left in my lungs began expanding again and thus I was able to resume blowing out air until reaching the surface. This action of vigorously expelling expanding air from the lungs is necessary to prevent an embolism, the effects of which can be permanently disabling or even fatal.

Buoyant exhaling ascent

Because *Alliance* and her class were never fitted with the single-man escape tower and associated equipment adopted for later submarine types, the escape method used in A-class boats was, in reality, similar to the

Davis Escape Equipment introduced before the Second World War. Colloquially known as free ascent, it was by far the quickest method of escape, although its singular disadvantage was that it placed considerable physical strain on the human body. Recognising this, the Royal Navy ceased to carry out pressurised submarine escape training in March 2009.

- The escape equipment used in *Alliance* consists of goggles, a strong nose clip and an inflatable stole worn much like a life jacket.
- Nobody can attempt to exit the submarine via the escape hatch until the following actions have been taken:
- The escape compartment (ie, either the fore or after torpedo compartment) has been fully shut off and is watertight from the rest of the submarine.
- The twill trunking has been rigged from the underside of the escape hatch.
- The escape compartment has been sufficiently flooded above the bottom rim of the twill trunking by opening the compartment flood valve.
- The escape compartment has been pressurised so that pressure is equal to that of the water pressure outside the submarine. This is monitored on the differential pressure gauge, which is connected to both sea and the escape compartment and is graduated 30–0–150psi. Indicating depth, it enables a check to be kept on differential pressure during flooding up. Pressure balance can be controlled by opening the 'air aiding' valve. Only when pressure is equalised can the escape hatch be opened.
- Each man is breathing from a BIBS (built-in breathing system) mask.

One by one each man exits the boat by taking a very deep breath, ducking underwater to enter the bottom of the escape twill trunking, and swimming up through the escape hatch. While ascending towards the surface, assisted by the buoyancy of the air-filled stole, the escapee must blow hard to exhale the expanding air in his lungs during ascent; failure to do so will result in the 'bends' and possibly an embolism.

As each man exits, the next moves towards the escape hatch, 'fleeting' from one BIBS mask to the next, the senior survivor being the 'last man out'. Once at the surface the survivors must group together and switch on their battery-operated stole lights for visibility.

ABOVE After escape compartment flood flap valve (right) and adjacent flood valve (left). *(Author)*

Bringing attention to a stricken submarine

This can be done by three means as follows:

- **Releasing the submarine indicator buoys**
Two are fitted, one under the foremost casing, the other under the after casing. The foremost is released from the fore torpedo room or adjacent accommodation space, the after buoy from the after torpedo room or adjacent motor room. When released,

LEFT After escape hatch, note twill trunking retaining pins. *(Author)*

each buoy floats to the surface, secured to the submarine by a 600ft length of wire. Once at the surface, a self-erecting aerial switches on a radio unit that automatically transmits a signal in a 10-minute cycle alerting coastal and naval services, their radar being used to track the submarine's position; a light on the buoy also flashes automatically, once per second for a period of 60 hours. Indicator buoys are painted 'International Orange' and have a ring of cats' eyes for night visibility. Plates bearing inscriptions are riveted to the top and read 'S.O.S. HMS Alliance FORWARD [or AFTER]. FINDER INFORM NAVY, COASTGUARD OR POLICE. DO NOT SECURE TO OR TOUCH.' These inscriptions are also written in French.

■ **Using the SSE (Submerged Signal Ejector)** This sends signal flares to the surface and is only used when the boat is known to be close inshore and when survivors know – from propeller noise – that rescue vessels are in close proximity on the surface. Differing flares are used for daylight or night.

■ **Creating noise** This is achieved simply by banging the hull with a hammer or large spanner at regular intervals. Noise carries well underwater and can attract attention from nearby vessels, whose crew will reply either by hammering or sending a sonar 'ping'. In addition, crew of surface vessels may, with great caution, use hand grenades to indicate their presence.

Escape procedure

The two main factors required for successful escape are:
■ Knowledge of the principles of escape and details of associated equipment.
■ Determination on the part of all escapers.

Unless a sunken submarine has been flooded throughout, one or other of the endmost compartments (ie, the after or fore torpedo room) is the least likely to be flooded. Therefore, all survivors need to make their way to the appropriate compartment and, once there, nominate one man as the 'senior survivor' who takes leadership of all personnel in the compartment, regardless of rank.

The senior survivor's first task is to open the red-taped lockers (two per escape compartment) sited near the escape hatch. These contain the escape instruction book (with information and procedures) together with two 'long breathing units', two pressure-tight flashlights and a wheel spanner. He assesses the number of survivors and their physical condition, and then consults the man-hour table in the escape instruction book for an approximate forecast of time available before escape. He will then, if possible, establish emergency lighting and rig torches to maintain morale, release the submarine indicator buoy and establish an unpolluted air supply.

When to escape

In reality this is a waiting game, as there is no point in escaping to the surface unless there is a good chance that help is close at hand. When surface ships locate a sunken submarine they drop 12 grenades to let the survivors know they are ready to pick them up. Then escape without delay. However, if a ship or ships appear to be in the vicinity, escape should not necessarily be delayed until the grenades are heard.

How long survivors wait will also depend on the condition of the air in their compartment and the degree and rate of any flooding. If there is flooding, immediate escape must be made – this is a 'rush escape'. In the absence of flooding, the factors governing the timing of escape are, first, the condition of the air and, second, the relationship between CO_2 and absolute pressure, as explained below.

Condition of air

In order to preserve the condition of the air in the compartment, air purification equipment must be started immediately after the accident using the following:

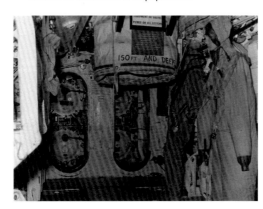

RIGHT Forward torpedo stowage compartment showing escape hatch with twill trunking lowered, and escape suits, 1978.

- **CO_2 absorption unit** Use four CO_2 canisters every two hours. Power permitting, it is preferable to run this electrically rather than use the alternative air drive, which will increase compartment pressure; as pressure increases, CO_2 levels become raised exponentially, causing greater adverse effect for the survivors.
- **O_2 generator** Burn one O_2 candle every two hours.

Air purification can be regarded as running at full efficiency if one CO_2 unit and O_2 generator is available for every 32 men; anything else is regarded as reduced efficiency. Once air purification can no longer be run at full efficiency and levels of CO_2 and O_2 become unable to sustain life, survivors must commence breathing from masks supplied from BIBS (built-in breathing system), if available; if not available, escape must commence.

If no air purification equipment is available, escape must commence.

Relationship between CO_2 and absolute pressure

When the time to escape is judged to be close, the senior survivor starts to take half-hourly readings of CO_2 percentage and compartment pressure (shown in bar on the absolute pressure gauge), and multiplies the two readings. Escape must commence when the multiplied figure reaches 4.5 or when the pressure gauge reaches 1.5 bar – whichever occurs first.

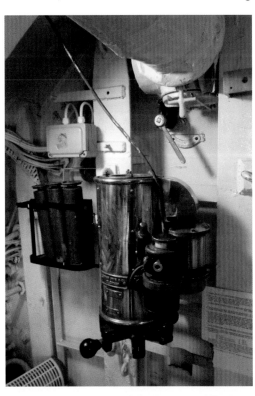

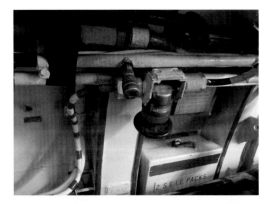

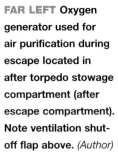

FAR LEFT Oxygen generator used for air purification during escape located in after torpedo stowage compartment (after escape compartment). Note ventilation shut-off flap above. *(Author)*

TOP LEFT Built-In Breathing System (BIBS) isolation valve and adjacent breathing unit (BU) bayonet connection. Note Submarine Escape Immersion Equipment (SEIE) locker below. *(Author)*

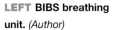

LEFT BIBS breathing unit. *(Author)*

FAR LEFT BIBS breathing unit hose inserted into system bayonet connection. *(Author)*

LEFT The absolute pressure gauge, important when evaluating when to escape with SEIE (locker below). *(Author)*

Chapter Ten

Restoring *Alliance*

─────●─────────────

Thanks to a £7 million programme of restoration and conservation at the Royal Navy Submarine Museum, HMS *Alliance* has been saved for future generations and is now in better condition than at any time since the end of her active life.

OPPOSITE Bow-on view of *Alliance* under the protective covering of the tented scaffold. Note how her bow frame stations have been numbered. *(Jonathan Falconer)*

167
RESTORING *ALLIANCE*

ABOVE A tired-looking *Alliance* in 1989 before her restoration.

When HMS *Alliance* was first put on public display in 1982 at the then newly opened Royal Navy Submarine Museum in Gosport, Hampshire, she was supported over seawater on concrete cradles next to the museum quayside. Although this provided a visually impressive and innovative means of display for 32 years, unfortunately it also created an inherent difficulty as the lack of accessibility to the external hull prevented regular maintenance to be undertaken to arrest continual corrosion to her bow, stern, fairing plates and keel.

Furthermore, with adverse conditions in the form of seawater always in close proximity and sea spray prevalent in windy conditions, the environment had taken its toll and parts of *Alliance*'s structure had deteriorated considerably, to the point where some areas were becoming unsafe. Breakdown of the original coatings, reactivation of chloride corrosion and the inception of electrolytic corrosion around *Alliance*'s many mixed-metal construction elements had occurred in the absence of the standard day-to-day maintenance that would have been carried out by the crew or depot ship when the boat was in commission. Gaping holes caused by aggressive rust over the years had also opened

LEFT Serious corrosion in the boat's stern free-flood section and after torpedo tubes …

BELOW … and in her bow.

up habitats for birds, especially feral pigeons, whose excrement had exacerbated corrosion still further in localised areas. The outer structure had corroded so badly that some parts were literally in danger of falling into the sea below.

In 2007 engineering consultants were appointed to carry out a feasibility study into the best long-term solution for the sustainable preservation of the vessel. The study evaluated a number of options including moving the submarine to a dry dock. However, the most cost-effective option was to reclaim the land under the submarine using a cofferdam and then carry out a full restoration using design and materials that matched the original build as closely as possible. A research paper about *Alliance*'s condition was presented to the ninth Maritime Heritage Conference, held at Baltimore, USA, in 2010. This paper outlined the steps that had been taken to survey the hull and the finding of the options appraisal. It also outlined the overall scope of works and potential solutions for hull restoration and the framework for a long-term maintenance programme. The survey work found the internal structure and fittings to be in good condition, consequently the internal work centred on the conservation of the existing fabric rather than the extensive replacement necessary for the external works.

Thanks to funding from major donors together with a £3.4 million grant from the UK's Heritage Lottery Fund, the Royal Navy Submarine Museum was able to embark on the restoration project. Experienced civil and maritime engineering consultants were invited to map out a comprehensive restoration programme with the following objectives:
- To restore the submarine with historical accuracy, conserving as much original fabric as possible.
- To make limited alterations to the vessel that will improve its long-term preservation.
- To implement a long-term maintenance strategy.
- To enhance physical and intellectual access to the submarine and to have greater community involvement.
- To extend the waterfront by reclaiming the land under the submarine.

ABOVE Under the fore casing: besides the general corrosion to the fore hydroplane operating mechanism, feral pigeons had contributed to the overall deterioration. *(Jonathan Falconer)*

A major problem with the restoration of large heritage vessels is finding enough experienced contractors to allow a competitive tender exercise to take place, and in Britain this is compounded by the fact that the number of repair yards has decreased. Furthermore, as it was not feasible to move the boat, it was necessary for the selected contractor to mobilise to Gosport to carry out the work. Before work could begin on *Alliance* herself, the project started with the construction of a cofferdam around the boat that would serve three purposes:
- Greater accessibility for continuous hull

BELOW Land was reclaimed underneath *Alliance* using a cofferdam and backfill.

maintenance, with provision for mechanical devices such as cherry pickers and mobile cranes.
■ Improved protection of the keel and lower hull from extreme tides.
■ Safe all-round public access for viewing the entire hull created by reclaiming land underneath the submarine, using a cofferdam and backfill.

External repair and restoration

Work on the submarine began in earnest in July 2012 with removal of the after hydroplanes to allow piling for the cofferdam by the civil engineering contractor. Once this

ABOVE A giant tent-like structure was built over the boat to enable contractors to work on her in all weathers.

RIGHT A protective island well above sea level was created underneath *Alliance*.
(Author)

RIGHT **The after free-flood casing was among dozens of decayed casing sections that were removed from the boat.** *(Jonathan Falconer)*

CENTRE **The port hydroplane and forward casing were typical examples of how the boat's external structure had deteriorated.** *(Jonathan Falconer)*

section was complete, the boat's after free-flood stern structure was removed. Although the guiding methodology was to preserve as much of the original fabric as possible, the stern was deemed to be too far gone to be worthwhile repairing. There was some delay in the construction of the cofferdam but once complete the submarine was encased in a sheeted three-storey scaffold. Every square metre of the vessel was then blasted with either high-pressure water or grit, including the interiors of all the external tanks running down both port and starboard sides of the boat. The blasting process allowed the contractors and the museum team to see the full extent of corrosion and therefore the real scope of repair work required. In total the amount of replacement plating rneeded was close to what had been estimated; however, the locations that proved the worst were not always where they were expected.

The blasting revealed some interesting hidden evidence such as etched draught marks that had disappeared under several layers of paint, something that had been obviated in service by welding on pieces of thick plate with marks carved deeply into them to serve as the upper draught marks.

As various corroded fairing plates and pieces of tank plating were removed, a variety of unexpected features came to light, many of which were not present on any extant drawing and therefore could only be

RIGHT **Ultrasound tests were carried out on the metal casing to determine the extent of corrosion. The recorded densities were noted in centimetres and painted on to the metal in yellow (the larger grey numbers correspond to the internal frame positions). This is the port bow area showing the housed anchor.** *(Jonathan Falconer)*

ABOVE Ultrasound tests were carried out on the ballast tank plating. *(Jonathan Falconer)*

BELOW The pressure hull with steel brackets supporting the superstructure casing. *(Jonathan Falconer)*

identified after further research. One find was an array of hydrophones (underwater sound receivers) along the flanks, an unexpected feature but one that is now commonplace in modern submarines.

Another was the discovery of a bronze cable reel. After some research it was identified as an experimental device for trailing an aerial cable fitted on the upper hull, an alteration that does not seem to have been recorded. Solving these little conundrums often requires the help of surviving crew members, and fortunately the museum had consulted widely with former A-class submariners.

Another surprise during the restoration was the fact that the boat was several inches shorter than she was supposed to be! As dismantling of the structure continued another unexpected feature came to light when it was found that the outer torpedo door mechanism had been disabled by the simple expedient of cutting a length out of the operating rods. A similar disabling method had been applied to the after hydroplanes, which also had sections removed from their hydraulic lines. While there is no evidence to indicate when these adaptations to *Alliance* occurred, it can be safely assumed that they were done during her last years in the water as a training vessel in order to prevent some careless trainee from sinking the boat. These discoveries finally put to rest any ideas about restoring these mechanisms to working order.

All engineering work was based on the best available information, which included an extensive (but not complete) set of steelwork drawings c.1945, supplied by BAE Systems from Vickers-Armstrong records at Barrow-in-Furness. Despite the usual occasional mismatch between what was drawn and what was found 'as built', these drawings proved very useful.

The process of restoration always provides inherent puzzles. One was to determine the exact shape of the missing hydroplane guards that had been removed several years earlier due to their advanced decay. Although they were meticulously measured and drawn at the time of their removal, the measured shape was quite unlike the available drawings; careful scrutiny of archive photographs, however, eventually revealed that these were 'post-streamlining' guards (fitted after the 1958–60 refit) that differed in both shape and construction.

When sections of the aluminium-alloy casing were being removed, the sheer scale of pigeon guano hindered work. Furthermore, the removal of casing sections revealed extensive decay of the steel brackets supporting the casing. The major concern arising from this was how to reinstate the casing without causing damage to the interior of the pressure hull and without moving away from historical authenticity. There were two specific practical problems to overcome: firstly, how to restore the steel brackets without damaging the pressure hull interior with excessive heat from welding; and secondly, how to deal with the mixed-metal corrosion that caused degradation at alloy/steel interfaces in the first place.

FAR LEFT General view of the corroded stem post and fore part of the hull. *(Jonathan Falconer)*

LEFT Stem post, and lower bow structure before restoration. Note fore end of torpedo tubes and their bow caps. *(Jonathan Falconer)*

FAR LEFT No 4 torpedo tube bow caps with No 2 above. *(Jonathan Falconer)*

LEFT No 1 torpedo tube bow cap amid corroded structure. *(Jonathan Falconer)*

FAR LEFT No 2 torpedo bow cap and corroded port bow frame structure above. *(Author)*

LEFT Bow section with Nos 1 and 3 torpedo tube bow doors after restoration. *(Author)*

RIGHT *Alliance*'s draught marks painted on her hull denoting 16, 17 and 18ft. *(Jonathan Falconer)*

FAR RIGHT Starboard 'A' bracket and propeller tail shaft bearing before restoration. *(Author)*

FAR RIGHT Exposed starboard diesel exhaust system. Also shown are the engine room hatch housing and Nos 3 and 4 main ballast tank access covers. *(Author)*

RIGHT Submarine Museum head curator Bob Mealings beside the starboard screw (propeller). *(Jonathan Falconer)*

FAR RIGHT After torpedo loading hatch. *(Author)*

RIGHT Towed array capstan. *(Jonathan Falconer)*

FAR RIGHT Welders at work below the refabricated stern section. *(Jonathan Falconer)*

Considerable work was undertaken at the fore end of the hull, particularly in repairing the housings around the fore torpedo tube bow caps. Here an integral 'web' structure with badly corroded fairing plates had to be removed and rebuilt with new metal welded into position. When steel repairs were complete, the hull was prepared for the paint coatings, the most important of which was the anti-corrosion undercoat. The conditions under which the paint could be applied had to be carefully managed so that it conformed to the manufacturer's requirements. If, for example, the paint was applied when the air was too moist the guarantees on the coating might be rendered invalid.

Painting went ahead on schedule in May 2013 and was finished successfully, bringing the external restoration phase almost to the point of completion by the end of July.

Internal conservation

Introduction

The internal conservation of *Alliance* can be divided into two areas of work, remedial and preventative. Carrying out the work was particularly challenging for the specialist conservation team because for much of their available working time – June 2013 to February 2014 – *Alliance* was still open to the public.

Remedial

Although the interior of the submarine was basically in sound condition and metal corrosion for the most part was largely superficial, there were still a great many surfaces and components suffering from tarnishing or decay. The surface corrosion not only needed to be dealt with for the long-term benefit of the historic fabric, but also because getting the submarine looking 'shipshape' rather than retired was a key part of the *Alliance* interpretation strategy.

The conservation contractors worked their way through the boat compartment by compartment. In the living spaces all the larger components (eg bunks and tables) were stripped out and individually cleaned and restored. The empty spaces were meticulously cleaned and variously polished and waxed

LEFT Seamen's mess and table. *(Author)*

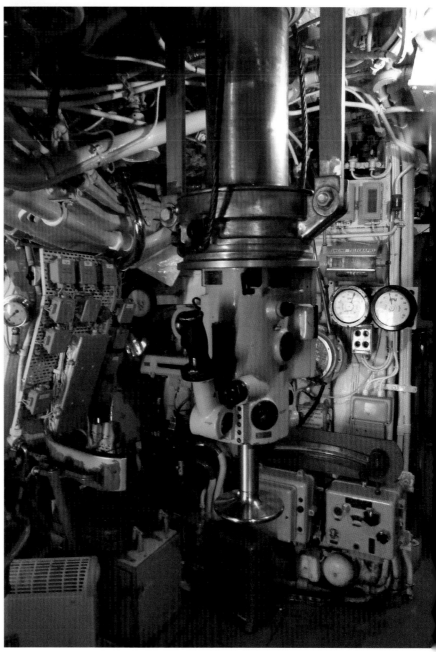

BELOW Inside the control room showing search periscope, hoist wire and pulley. *(Author)*

depending on the material finish. The remedial work made good many years of inadequate care and maintenance with the result that the interior looked shipshape and operational.

Preventative

Early on in the planning stage for the project it was anticipated that a new comprehensive air-conditioning system would be introduced to service every area within the submarine's pressure hull. The intention was to have an air-conditioning system that could maintain the relative humidity at approximately 50–60 per cent all year round to prevent the return of surface corrosion. The scheme would also provide chilled air in the summer months. Various systems were investigated and costed, but it became clear that the spatial dynamics of the vessel's interior would make it very difficult to achieve even the relatively modest parameters required all year round. Furthermore all the systems considered were very expensive

to install and would generate substantial running costs while not actually delivering a perfect atmosphere.

The cost benefit equation did not seem to justify the level of capital investment required so the museum opted for a much more simplistic approach. Low-cost 'conservation heating' was installed in the form of electric tubular heaters in every compartment to raise winter temperatures sufficiently to ensure that very high relative humidity levels were avoided. For fresh air and cooling new fans were installed and connected to the original main ventilation trunk that runs through almost every space inside the submarine's pressure hull.

The submarine now has in place a simple low-cost system that prevents the damaging extremes of excessive cold, heat and humidity. The remedial conservation works also make a major contribution because surfaces that were once exposed to corrosion are now sealed to protect them from future decay.

BELOW **Restoration completed and reopened to the public.** *(Jonathan Falconer)*

Appendix 1

List of sources and further reading

PRIMARY SOURCES

Royal Navy Submarine Museum, A2003/63

BR.1963 (3) 1950, Handbook for Submarines of the 'A' Class, part 3 – The Ballast Pump.

BR.1963 (7) 1950, Handbook for Submarines of the 'A' Class, part 7 – Hydroplane Gear.

CB. (R) 4549 (2) 1951, Handbook for Submarines of the 'A' Class, part 2 – Tank and Systems.

BR.1963 (11) 1951, Handbook for Submarines of the 'A' Class, part 11 – Periscopes.

BR.1963 (13) 1951, Handbook for Submarines of the 'A' Class, part 13 – Submarine Log Mark 4 and Projector Binnacle ACO Mark II.

BR.1963 (8) 1952, Handbook for Submarines of the 'A' Class, part 8 – Main Engine Associated Equipment.

BR.1963 (6) 1957, Handbook for Submarines of the 'A' Class, part 6 – Steering Gear.

BR.1963 (4) 1958 Handbook for Submarines of the 'A' Class, part 4 – Auxiliary Machinery.

BR.1963 (9A) 1959. Handbook for Submarines of the 'A' Class, part 9A – Vickers Main Engine.

BR.1963 (9) 1959. Handbook for Submarines of the 'A' Class, part 9 – Admiralty Main Engine (used for comparison notes only).

Royal Navy Submarine Museum, A2007/1016

BR.1963. Handbook for Submarines of the 'A' Class (includes parts 3, 4, 6, 7, 8, 9, 9A, 11 and 13).

Royal Navy Submarine Museum, A2007/782

Captain's Ship's Book Submarines, Vol 1 HMS Alliance, date-stamped 1 April 1963 to 14 July 1972.

HMS Excellent Museum

BR.205/47, Drill for 4inch Mark XII, XII* and XXII on SI Mountings (Submarines), 1947.

BR.968 1925, Handbook for 4-inch QF Mark XII gun on SI Mounting.

Hampshire County Record Office, Winchester

BR.1845 Handbook for 4-inch QF Mark XXIII gun on 4-inch S2 Mounting, 1949 (Priddy's Hard Collections).

BR.1845 Handbook for the 4-inch QF Mark 23 gun on a 4-inch S2 MOD 1 mounting, technical description. March 1969. (Royal Navy Submarine Museum Collections).

Miscellaneous papers and documents

HMS Artemis, *Lessons Learnt* (author's collection).

Davies, W.B. and Mealings R., *The Submarine Alliance*, 9th Maritime Heritage Conference, Baltimore, 2010.

Davies, W.B. (5201 Consultancy Limited UK), *Restoration of the Submarine* Alliance, Historic Ships Conference RINA 2013.

RNSM 2006.88.240 – Crest of HMS Alliance with motto.

BELOW HMS *Alliance* ship's crest. The badge represents the alliance of the fighting services. On a 'field blue' the badge shows 'in front of two wings conjoined white two swords in saltire also white pommels and hilts gold surmounted by an anchor also gold'.

ABOVE The dolphins badge is awarded to all Royal Navy submariners on completion of their training.

SECONDARY SOURCES

Admiralty Committee, *Naval Engineering Practice Vols 1 & 2* (HMSO, 1959)

Akerman, P., *Encyclopedia of British Submarines 1901 to 1955* (London, Periscope Publishing, 2002)

Blamey, J., *A Submariner's Story: the Memoirs of a Submarine Engineer in Peace and War* (London, Periscope Publishing, 2002)

Booth, T., *Thetis Down: the Slow Death of a Submarine* (Barnsley, Pen & Sword, 2008)

Bowers, P., *The Garrett Enigma and the early Submarine Pioneers* (Shrewsbury, Airlife Publishing, 1999)

Branfill-Cooke, R., *X.I – the Royal Navy's Mystery Submarine* (Barnsley, Pen & Sword, 2013)

Clayton, T., *Sea Wolves: The Extraordinary Story of Britain's WW2 Submarines* (Abacus, 2012)

Compton-Hall, R., *Submarine Boats* (Cambridge University Press, 1983)

Compton-Hall, R., *The Submarine Boats: the Beginnings of Submarine Warfare* (London, Windward, 1983)

Compton-Hall, R., *The Submarine Pioneers* (Stroud, Sutton Publishing, 1999)

Coote, J. Capt. RN, *Submariner* (Barnsley, Pen & Sword, 2014)

Cook, G., *Silent Marauders: British Submarines in the two World Wars* (London, Hart-Davis MacGibbon, 1976)

Dickison, A.P., *Crash Dive: in Action with HMS Safari 1942 to 1943* (Stroud, Sutton Publishing, 1999)

Dornan, P., *Diving Stations: the Story of Captain George Hunt and the Ulto* (Barnsley, Pen & Sword, 2010)

Gallop, A., *Subsmash: the Mysterious Disappearance of HM Submarine Affray* (Stroud, The History Press, 2011)

Hart, S.B., *Submarine Upholder* (Stroud, Amberley Publishing, 2009)

Lambert, J. and Hill, D., *The Anatomy of the Submarine Alliance* (London, Conway, 1986)

McGeorge, H.D., (Ed) *Marine Auxiliary Machinery 2nd Edition* (Committee of Marine Engineers, London, 1955)

Murphy, W.S., *Father of the Submarine: the Life of the Revd. George Garrett Pasha* (London, William Kimber, 1987)

Newton, R.N. RCNC, MINA, *Practical Construction of Warships* (London, Longmans, 1964)

Kemp, P.J., *The T Class Submarine: the Classic British Design* (London, Arms & Armour Press, 1990)

Preston, A., *The Royal Navy Submarine Service: a Centennial History* (London, Conway, 2001)

Shelford, W.O. RN (Rtd), *Subsunk: the Story of Submarine Escape* (London, Harrap, 1960)

Young, E., *One of Our Submarines* (London, Penguin, 1954)

Appendix 2

Glossary of terms

After ends After torpedo stowage compartment inclusive of after torpedo tubes.

AIV Automatic Inboard Venting tank. Once torpedo is fired this provides appropriate compensating water for weight change, maintaining longitudinal balance (submarine trim).

Amp tramp An electrician (also a 'Greenie').

Donk shop Colloquialism for engine room; sometimes coupled with 'SPO' to make 'donk shop Stoker PO'.

Fish Colloquialism for torpedo.

Fore endie Colloquialism for fore torpedo man.

Fore ends Fore torpedo space and tubes afore 26 bulkhead; also, colloquially, the fore torpedo stowage compartment including the fore torpedo space.

Fruit machine Torpedo control calculator.

Grocer Colloquialism for the coxswain (responsible for provisioning from various sources).

Hooky Nickname for a leading rating distinguished by insignia of a single fouled anchor on his left sleeve, the anchor colloquially called a 'hook'.

Outside wrecker Colloquialism originates from days when a submarine's boarding party included one ERA (Engine Room Artificer) whose role was to scuttle a 'wreck' – a captured vessel. This task invariably fell to the 'Clanky' ERA or Mechanician responsible for the entire fore mechanical hull systems and auxiliary machinery 'outside' the engine room and main propulsion units.

Roughers Heavy seas that cause a boat to pitch and roll.

SPO Stoker PO (Petty Officer).

Trot One or more submarines tied alongside each other against the jetty or submarine depot ship.

Trot sentry A rating detailed to guard over a trot of boats, trotting from one to another.

Watching the bubble Action of planesmen watching inclinometer when maintaining trim.

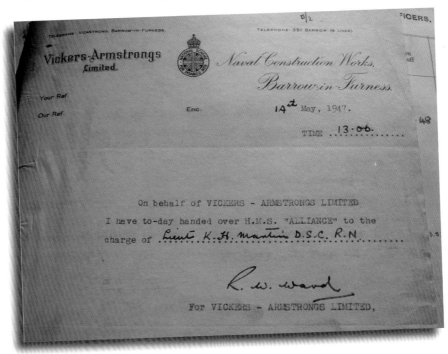

RIGHT Official handover document for HMS *Alliance* to the Royal Navy on 14 May 1947.

Appendix 3

A-class submarine specifications

Length, overall	281ft 5in (85.78m)
Length, between perpendiculars	249ft 3in (75.97m)
Maximum breadth	22ft 3in (6.78m)
Pressure hull diameter	16ft 0in (4.88m)
Draught	17ft 0in (5.18m)
Displacement, standard	1,120 tons
Displacement, normal surfaced	1,385 tons
Displacement, submerged	1,620 tons
Operating depth	500ft (c.150m)
Maximum deep diving depth (DDD)	750ft (c.230m)
Deep diving depth test pressure (DDDTP)	375psi (25.5 bar)
Number of watertight compartments[1]	Seven
Pressure hull plating	0.875in (22.3mm) thick, reducing to 0.75in (19.1mm)
Armament, torpedo tubes	10, 21in
Armament, torpedoes	20, 21in Mk VIII
Armament, deck	One quick-firing 4in gun
Armament, conning tower	One 20mm Oerlikon cannon, three .303 Vickers GO machine guns
Armament, other	26 mines[2]
Main machinery drive	Two Vickers eight-cylinder diesel engines, 2,150bhp each
Electrical drive	Two English Electric motors, 625bhp each
Propeller shafts	Two
Maximum speed, surfaced	18.5kts
Maximum speed, submerged	8.0kts
Endurance, surfaced	10,500 nautical miles at 11kts
Endurance, submerged	16 nautical miles at 8kts, 90 nautical miles at 3kts
Fuel capacity (diesel)	35,616 gallons (159 tons)
Estimated fuel consumption	3 miles per gallon
Complement	61

Notes
[1] Includes the Captain's segregated cabin located within the conning tower structure.
[2] Most submarines were capable of laying mines.

Appendix 4

A-class submarine weights and capacities

Hull	571.03 tons	–
Main machinery	140.00 tons	–
Main motors	37.00 tons	–
Batteries	118.90 tons	–
Torpedo armament	70.00 tons	–
Gun armament	13.30 tons	–
Electronic gear	11.10 tons	–
Air compressors	3.14 tons	–
Cooling plant	4.00 tons	–
Lubricating oil	18.70 tons	4,188.8 gallons
Fuel oil (diesel)	165.20 tons	37,004.8 gallons
Fresh and distilled water	27.80 tons	6,227.2 gallons
Compensating water	23.00 tons	5,152.0 gallons
Trimming water	10.00 tons	2,240.0 gallons
Crew and their effects	5.60 tons	–
Provisions and stores	12.00 tons	–
Air reservoirs	7.50 tons	–
Hull, spare gear	1.00 tons	–
Ballast, etc	117.70 tons	–
Total	1,356.97 tons	–

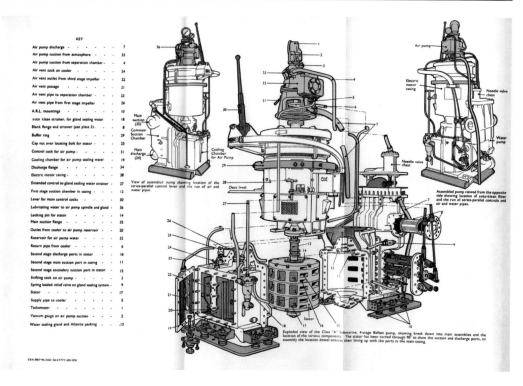

RIGHT A-class submarine 4-stage ballast pump.

Appendix 5

Preserved submarines open to visitors in the UK

The following is a listing of the submarines open to visitors in the UK, six of the nine being at the Royal Navy Submarine Museum. Submarines open to the public worldwide are listed in the table in Appendix 6.

HMS *Alliance* (P417/S67) Amphion-class submarine. Launched and commissioned in 1947, she is the only survivor from the Royal Navy's A-class submarines, specially designed for long-range duties in the Pacific Ocean. In 1981 she began service as a museum and memorial to lost British submariners, at the Royal Navy Submarine Museum, Gosport, Hampshire.

HMS *Holland 1* Launched in 1901, the *Holland 1* was Britain's first commissioned submarine. She was decommissioned in 1913 and sank while being towed to the breaker's yard. Her wreck was located in 1981 and recovered a year later. Now restored and conserved, she is displayed at the Royal Navy Submarine Museum, Gosport, Hampshire.

X24 The only remaining British X-class midget submarine from the Second World War. Launched in 1943, she was designed for commando missions, and is powered by the same diesel engine used by London buses. Displayed at the Royal Navy Submarine Museum, Gosport, Hampshire.

Biber 105 German midget submarine. Displayed at the Royal Navy Submarine Museum, Gosport, Hampshire.

HMS *E17* conning tower Preserved as a monument at the Royal Navy Submarine Museum, Gosport, Hampshire.

LR3 (submersible) Ocean survey and repair submersible, built 1982. Displayed at the Royal Navy Submarine Museum, Gosport, Hampshire.

HMS *Ocelot* (S17) Launched in 1962, HMS *Ocelot* is a diesel/electric submarine and the last built at Chatham Dockyard; she served until 1991. Now open to the public at the Historic Dockyard in Chatham, Kent.

U-534 A Type IXC/40 U-boat built in 1942. She is famous for refusing the surrender orders issued by German Admiral Dönitz in May 1945 but RAF aircraft located and sank her. The wreck was raised in 1993 and is now displayed in sections at the Woodside Ferry Terminal, Birkenhead.

Biber 90 German midget submarine, displayed at the Imperial War Museum, London.

LEFT German Type IXC/40 U-boat U-534 is a popular tourist attraction on Merseyside. *(Merseytravel)*

Appendix 6

Preserved submarines open to visitors worldwide

	Class/Type	Date	Location
Australia			
M-14/M-21 Japanese midget submarine	Ko-hyoteki	1942	Canberra
HMAS *Ovens*	Oberon	4 December 1967	Fremantle, Western Australia
HMAS *Otway*	Oberon	1966	Holbrook, NSW
HMAS *Onslow*	Oberon	3 December 1968	Sydney, NSW
M-21 conning tower (Japanese)	Ko-hyoteki	1942	Sydney, NSW
Belgium			
B-821 (sometimes called B-143)	Foxtrot	1960	Zeebrugge
Canada			
HMCS *Ojibwa* (S73) [originally HMS *Onyx*]	Oberon	29 September 1965	Port Burwell
HMCS *Onondaga* (S72)	Oberon	25 September 1965	Quebec
Ben Franklin deep ocean explorer	RV	1982	Vancouver
Croatia			
CB-20	Midget	WWII	Zagreb
Estonia			
ML *Lembit*	Kalev	1936	Tallinn
Finland			
Vesikko	–	1933	Helsinki
France			
Argonaute (S-636)	Aréthuse	9 November 1957	Paris
Flore (S-645)	Daphné	1961	Lorient
Espadon (S-637)	Narval	1958	Saint-Nazaire
Alose	Naïade	1904	Marseille
Redoutable (S-611)	SSBN	29 March 1967	Cherbourg
Germany			
U1 (first German military submarine)	U-boat	1906	Munich
U-995 (later in Norwegian navy as *Kaura*)	Type VII VIIC/41 U-boat	1943	Laboe
S-18	Oberon	1963	Sassnitz
U-434	Tango (Projekt 641b)	1960s	Hamburg
U-461	Juliett (Projekt 651)	1960s	Peenemünde
U-9 German midget 'Seehund'	Type XXVII B5 (type 127)	1944	Speyer Technical Museum
U-10	Type UB I U-boat	1915	Wilhelmshaven
U-11	Type UB I U-boat	1915	Fehmarn
U-2540 (*Wilhelm Bauer*)	Type XXI U-boat	1945	Bremerhaven
B-515	Tango	1976	Hamburg
India			
INS *Kursura* (S20)	Foxtrot	1958?	Visakhapatnam
INS *Vela* (S40)	Foxtrot	1973?	Tamil Nadu

LEFT German Type XXI U-boat U2540 is on display at the Deutsches Schiffahrtmuseum at Bremerhaven, where she is open to the public.

	Class/Type	Date	Location
Indonesia			
KRI *Pasopati* (410)	Whiskey	1950s	Surabaya, East Java
Israel			
INS *Gal*	Gal	1975	Haifa
Italy			
Enrico Toti (S-506)	Toti	1967	Milan
Nazario Sauro (S-518)	Sauro	1976	Genoa
Japan			
JDS (MSDF) *Akishio*	Yushio	1985	Kure
HA-18 midget submarine, used Pearl Harbor	Type 'A'	1941	Etajima
The Netherlands			
HNLMS *Tonijn* (S-805)	Potvis	1965	Den Helder Museum
Peru			
BAP *Abtao*	Lobo	1954	Callao
Russia			
B-413	Foxtrot	1969	Kaliningrad
B-396	Tango	1980	Moscow
B-307	Tango	1972	Togliatti
M-261	Quebec	–	Krasnodar
D-2 Narodovolets submarine	Dekabrist	1931	St Petersburg
C189 (S-189)	Whiskey Projekt 613	1954	St Petersburg
S-56	Strednyaya	WWII	Vladivostok
B-440	Foxtrot	c.1956	Vytegra

	Class/Type	Date	Location
South Africa			
SAS *Assegaai* (formerly *Johanna van der Merwe*)	Daphné	2 July 1970	Simons Town
Sweden			
S-363	Whiskey	16 November 1956	Karlskrona
Turkey			
CG *Uluçalireis* (formerly USS *Thornback*, SS-418)	Tench	7 July 1944	Istanbul
UB-46 (Imperial German Navy, used at Dardanelles)	Type UB II U-boat	WWI	Çanakkale
USA			
USS *Drum* (SS-228)	Gato	12 May 1941	Mobile, Alabama
U-505 (German)	Mk IX-C	24 May 1941	Chicago, Illinois
USS *Silversides* (SS-236)	Gato	26 August 1941	Muskegon, Minnesota
USS *Bowfin* (SS-287)	Balao	7 December 1942	Honolulu, Hawaii
USS *Croaker* (SS-246)	Gato	19 December 1942	Buffalo, New York
USS *Cod* (SS-224)	Gato	21 March 1943	Cleveland, Ohio
USS *Batfish* (SS-310)	Balao	6 May 1943	Muskogee, Oklahoma
USS *Pampanito* (SS-383)	Balao	12 July 1943	San Francisco, California
USS *Ling* (SS-297)	Balao	15 August 1943	Hackensack, New Jersey
USS *Lionfish* (SS-298)	Balao	7 November 1943	Fall River, Massachusetts
USS *Cavalla* (SS-244)	Gato	14 November 1943	Galveston, Texas
USS *Cobia* (SS-245)	Gato	28 November 1943	Manitowoc, Wisconsin
USS *Razorback* (SS-394)	Balao	27 January 1944	North Little Rock, Arizona
USS *Becuna* (SS-319)	Balao	30 January 1944	Philadelphia, Pennsylvania
USS *Torsk* (SS-423)	Tench	6 September 1944	Baltimore, Maryland
USS *Requin* (SS-481)	Tench	1 January 1945	Pittsburgh, Pennsylvania
USS *Clamagore* (SS-343)	Balao	25 February 1945	Charleston, South Carolina
USS *Albacore* (AGSS-569)	Albacore	August 1953	Portsmouth, New Hampshire
USS *Marlin* (SST-2)	T-1	14 October 1953	Omaha, Nebraska
USS *Nautilus* (SSN-571)	Nautilus	12 January 1954	Groton, Connecticut
USS *Growler* (SSG-577)	Grayback	5 April 1958	Brooklyn, New York
USS *Blueback* (SS-581)	Barbel	16 May 1959	Portland, Oregon
B-39 (Russian)	Foxtrot	15 April 1967	San Diego, California
USS *Dolphin* (AGSS-555)	Dolphin	8 June 1968	San Diego, California
B-427 (Russian)	Foxtrot	22 June 1971	Long Beach, California

Note With the exception of the French SSBN (Submerged Ship Ballistic Nuclear) *Redoutable*, which is nuclear-powered, all other boats listed are powered by diesel engines.

Index

A-class submarines and HMS *Alliance*
Accommodation and messes 8, 24, 57, 62, 152-157, 175
    bathrooms 60
    galley 59, 153-154
    sick berth 156
    toilets and washrooms 60, 156
    wardroom 57-58, 62, 155
Aerials 55
Air-conditioning 24, 76-78, 176
Air purification systems 78-80, 164-165
Auxiliary machinery and equipment 73

Baptism on board 35
Batteries 27, 103-105, 141, 156
    charging 104-105, 141
    explosion 41
    fuse panel 57
Bow, caps and doors 5, 33-34, 167-168, 173
Bridge 32-34, 36, 45, 55, 151
Build programme 24-25
Bulkheads 56, 58, 83

Capstans and anchor gear 132-133
Conning tower fin 8, 28, 36, 54-55, 137, 144
Control room 35, 55, 57, 59, 123, 129, 175
    clinometers 58-59, 138-139
    compass 55-57
    range finder 129
    telegraphs 57-58, 61, 93, 95, 137, 157
Crest plaque 34
Crew 24, 34, 147-157
    ratings 39, 41
    ship's company 148, 150
Crew members
    Baker, Alan 152, 154, 156-157
    Coles, George 157
    Daish, Mick 154, 157
    Draper, Mike 157
    Forster, Leading Seaman 151
    Handyside, Bill 152, 154-157
    Moore, Leading Cook Kevin 'Pony' 153
    Onions, Jim 153-154
    Pierce, Peter 154
    Potts, Ron 157
    Thomson, PO 151
    Whittaker, Lewis 154
Decommissioning 7, 41
Design and construction 24-26, 51, 54-55

Electrical system 103-105
Engine and motor room 37, 54-55, 61, 83-84, 92, 174
    fire 39

Exhaust system 61, 83, 87, 92, 174

Faring plates 168
Fate of the A-class 29
Fuel system 87, 92, 101-103
    centrifugal separators 101-103

Groundings 35, 38
Gun access tower 57

Hatches 54-55, 62, 144, 174
    escape 55, 163
    torpedo loading 54, 56
High-pressure air system 69-71
Hitting the seabed 41
Hull 8, 26, 28, 36, 173-174
    internal 55
    pressure 48, 51, 54-55, 177
Hydraulic systems 27, 123
    main telemotor system 71-73, 94
Hydroplanes 8, 30, 51, 57, 121-122, 124-125, 139, 169, 172

Keel 26, 54, 168

Launches 25-26, 30

Masts 27, 140

Name 7

Oily bilge system 81

Paint schemes 38, 175
Periscopes 9, 13, 125-131, 140, 144
    attack 26, 57, 59, 121, 126, 143
    search 26, 57, 59, 121, 126, 129, 175
Preservation 8-9
Propulsion 83-105
    diesel engines 61, 83-92
        lubricating system 88-89
        maintenance 91
    electric motors 61, 84, 92-98
    propellers and shafts 93, 96-98, 174
    superchargers 87, 92

Radar mast and systems 28, 60, 131-132
Range 24, 27, 83
Rebuild 36
Refitting 28-29, 32-36, 38-39, 92
Refrigeration system 76
Restoration 5, 7-9, 167-177
    conservation heating 177
    corrosion 168-169, 171, 175-177
    internal conservation 175-176

Rudder 51, 58, 121

Service history 26-28, 30-41
Snort mast and system 26-28, 30-31, 33, 59, 78, 98-100, 140-142
Sonar systems (ASDIC) 54, 132
Specifications 181
Speed 24, 27, 83-84
Steering gear and control 57-58, 122-125, 138
Stern 8, 168, 173
    fin and skeg 64, 125
Strong backs 55

Tank blowing and venting systems 66-69, 135, 138
Tanks
    bow buoyancy 62, 65
    compensating 51, 65-66, 73
    external 47, 62-64
        fuel 64-65, 101
    internal 46
    main ballast 48-51, 54, 57-58, 62-64, 136, 172, 174
    Q 65, 145
    sewage 60
    trim 51, 62, 73
    water and distillate 60, 80
Torpedo control calculator (TCC) 140
Torpedo room and tubes 55-56, 61-62, 109-110, 133, 143, 168, 172-173
Trim pump and system 74-75

Ventilation system 41, 141
Valves 50, 63
Voice pipe 55

Water systems 75-76, 80
Weapons 107-119
    guns and ammunition 114-119
    torpedoes 108-114
Weapons control centre 55
Weights and capacities 182
Wheelhouse 55
Workshop facilities 153

General
Accidents and losses 15, 18-19, 29-30, 34, 41, 159-161
    bringing attention to a stricken submarine 163-164
Admiralty 12, 17, 43
Affray Submarine Memorial Trust 160
American Electric Boat Co. 12

Anatomy and equipment
   boiler rooms 15
   bulkheads 14, 56, 58, 83
   funnels 14-15
   hulls 8, 15, 21, 26, 28, 36, 51, 54-55, 173-174, 177
   hydraulic systems 14-15, 27, 123
   sensors and sonar 16, 28, 36, 54, 132
   wireless transmitters and aerials 13, 36, 55
Armament
   anti-aircraft guns 118-119
   deck guns 15-17, 20, 24, 28, 32, 36, 38, 114-119, 144
   mines 119
   torpedoes 15-17, 20, 24, 55, 62, 108-114, 142-144
   torpedo tubes 15-17, 19-20, 24, 26, 28, 33-34, 36, 62, 109-110, 160
Attenborough, Richard 33

Bacon, Capt Reginald 12-13
BAE Systems 12-13, 172
Baker, Roy Ward 33
Bauer, Wilhelm 160
Buoyancy 13, 48, 65
Burdett, Lt A.J. D'Arcy 33

Cable, Frank 12
Cammell Laird 26
Chalmers, Lt Cdr A.T. 34
Chatham Dockyard 26, 183
Churchill, PM Winston 24
Clark, Lt Gen Mark W. 17
Clarke, Lt B.W.M. 33
Clutterback, Lt Cdr H.R. 34
Cold War 9, 23, 28, 36, 152
Colston, Lt B.A. 34
Crews 20, 147-157
   commissioned officers 150
   coxswain 156
   engine room artificer (ERA) 152-153, 157
   medical staff 152
   senior ratings 151-152
   training 148

Das Boot film 57
Davis Escape Equipment 162
Devonport Dockyard 24, 26
Diving 136-137, 144-145, 161
   to avoid depth charges 145
   radio signal 161
D nitz, Admiral Karl 183

Eddystone Lighthouse 13
Escape procedures 133, 159-165
   BIBS (built-in breathing system) 163, 165
   free ascent 162-163
   senior survivor 164
Exercises
   Britex 72 41

FOMEASWEX III 37
FOTEX 65 38
Goldfish 41
Lime Jug 40
Running Scrap 40
Sarder 41
Springboard 35
Strong Express 41
Sunny Seas 41

Faslane submarine base 43
First World War 13-16
Fisher, Sir John Arbuthnot 'Jacky' 12
Forsyth, Lt Cdr R.S. 39
French Navy 37

Greek Navy 19

Hammer, Lt Cdr C.H. 34
Heritage Lottery Fund 8, 169
HM Queen Elizabeth II 34, 43
HMS *Dolphin*, Gosport 7, 9, 18, 41, 148
   HMS *Affray* Memorial Service 23, 34
HMS *Ganges* 38
Holland, John P. 12
Hughes, Gwyneth, Harry and Minna 35

Imperial War Museum, London 183
Inter-war years 16-18
Israeli Navy 18

Joint Maritime Warfare School 35

Living in the boat 152-157
   clothing 37, 155
   cooking and food storage 153
   medical arrangements 156
   mealtimes 154
   personal hygiene 76, 155-156
   recreation 156-157
   rum 157
   shaving 32
   sleeping berths 154-155, 157
   smoking 31, 156

Maritime Heritage Conference 169
Martin, Lt Cdr J.D. 32
Martin, Lt K.H. 31
McShane, Ian 9
Merchant ships
   Almdijk 161
   Conte Rosso, liner 21
   Divina, tanker 18, 161
   SS *America* 129
Mills, Sir John 33, 57
Ministry of Defence 9
Morning Departure (*Operation Disaster*) film 33
Mountbatten of Burma, Earl 43

Nixon-Eckersall, Lt Cdr C.A.B. 39
Nobes, Lt Cdr B.I. 38

On the Beach film 28
Operation Barclay 17
Operation Flagpole 17
Operation Mincemeat 17
Operation Torch 17

Parnall Peto aircraft 15
Pearl Harbor 24
Peglar, Padre 35
Pender-Cudlip, Lt A.D.E. 40
Pennant numbers 25, 36, 152
Performance
   diving depth 14, 17, 24
   surface speed 14, 20
   submerged speed and endurance 16, 20
Pogson, Lt Cdr A.G.A. 37
Portsmouth 12, 15
Preserved submarines 8-9, 183-186
Propulsion systems
   battery-electric 12-14
   diesel-electric 7, 15-16, 20-21, 24, 43
   electric motors 24, 61
   high-test peroxide fuel 42-43
   internal combustion 12
      diesel engines 13, 17, 20, 24, 61, 83
      petrol engines 13
   petrol-electric 21
   nuclear 7, 41-43
   steam/steam turbine 12, 14, 21, 43
Purdy, Lt Cdr John P.A. 38

Redshaw, Sir Leonard and Mrs 30
Rickover, Admiral Hyman George 43
Royal Australian Navy 14, 38
Royal Canadian Navy (RCN) 35
   1st and 3rd Escort Squadrons 34
Royal Marine commandos 37-38
Royal Navy and RFA ships and tugs
   'Dreadnoughts' 12
   Dryad-class 12
   HMS *Agincourt* 29
   HMRT *Alliance* 7
   HMS *Alliance* (*Alliante*) 7
   HMS *Ark Royal* 40
   HMS *Fearless* 161
   HMS *Foudroyant* 23
   HMS *Hazard* 12
   HMS *Loch Insh* 29
   HMS *Manxman* 40
   HMS *Trincomalee* 23
   RMAS *Goldeney* 39
   RMAS *Kinloss* 39
   RMAS *Samson* 39
   TS *Royalist* 41
Royal Navy Submarine Museum, Gosport 5, 7-9, 13, 23, 162, 167-169, 183
   Mealings, Bob, Head Curator 174
   Munns, Chris, Director 9
Royal Navy Submarine School, Torpoint 8